DESIGN FOR IMPACT

Your Guide to Designing Effective Product Experiments

ERIN WEIGEL

Foreword by Stuart Frisby and Lukas Vermeer

TWO WAVES
BOOKS

TWO WAVES BOOKS
NEW YORK, NY, USA

"Erin Weigel offers practical and valuable insights to the world of experimentation. Covering process, prioritization, optimizing workflows, and statistical concepts, all intertwined with splashes of humor—it's a welcome dose of practicality for experimentation I wish I had been provided with when starting in the industry. Buy this book if you want to jumpstart the experimentation practices and principles in your workplace."

—Preston Daniel
Associate Product Manager, Personalisation, Endeavour Group

"*Design for Impact* is a masterpiece; it is modern product discovery from a designer's perspective. Erin brings an iterative and experimental approach that enables designers and gives them the knowledge to create better products!"

—Phil Hornby
Product Coach, for product people

"It's a valuable tool for students who want to better understand the experimentation culture within tech companies. An insider's perspective can help them determine whether their skills and aspirations are a good fit and connect what they already know with the demands of the field."

—Dr. Natalia Sánchez Querubín
Assistant Professor in Digital Culture at the University of Amsterdam

"Fundamental for improving online products and data-driven decision-making. Plus, Weigel's writing is light, funny, and supported by tons of helpful visuals."

— Lucia van den Brink
Founder, Increase Conversion Rate.com and Women in Experimentation

"Weigel's book is a beacon for product teams, offering the Conversion Design model and a comprehensive toolkit for refined experimentation processes. It's the cornerstone of building an Experimentation Culture."

—Adam Thomas
Principal, *Approaching One*

"Weigel turns the art of design into a science experiment you'll actually want to do in *Design for Impact*. Forget gut feelings—this book is all about the thrill of evidence, the beauty of data, and making decisions that pack a punch. A must-read if you're into winning arguments with hard facts and transforming your ideas into undeniable success stories."

—Janna Bastow
CEO and Co-founder, ProdPad

"This is a must-read book for anyone trying to get into experimentation. It's basically an Experimentation Bible. Erin was able to take incredibly challenging concepts and distill them into easy-to-digest readings, which are actually interesting to read! The crowdsourcing of building this book shouldn't be overlooked either—this isn't Erin declaring what to do. Rather, it's the sharpest minds in the CRO community discussing their thoughts and opinions, and Erin has distilled those thoughts into a book to make it accessible for everyone!"

—Shiva Manjunath
Host and Founder of the experimentation podcast From A to B

"Wow, *Design for Impact* really hit the mark for me! As someone deep into experimentation, I found the math sections super insightful. What stood out was how Erin breaks down complex math into something anyone can get. Whether you're new to the game or have been around the block a few times, this book has something for you. Big recommend for anyone looking to make sense of data and drive real change. Kudos to Erin for making this gem accessible and fun to dive into!"

—Jonas Alves
CEO, ABsmartly

Design for Impact
Your Guide to Designing Effective Product Experiments
By Erin Weigel

Two Waves Books
an imprint of Rosenfeld Media, LLC
125 Maiden Lane
New York, New York 10038
USA

On the Web: www.rosenfeldmedia.com
Please send errata to: errata@rosenfeldmedia.com

Publisher: Louis Rosenfeld
Managing Editor: Marta Justak
Interior Layout: Danielle Foster
Cover Design: Heads of State
Illustrator: Erin Weigel with support from
Alexis Oh, Dewi Williams,
Serbastian Tan, and Martin Pavlovic
Indexer: Marilyn Augst
Proofreader: Sue Boshers

© 2024 Erin Weigel
All Rights Reserved
ISBN: 1- 959029-37-1
ISBN 13: 978-1-959029-37-3
LCCN: 2024930249

To my daughter, Hedda, for letting me read my book to you as a bedtime story. I loved how invested you were in the tea tasting experiment and how we drew bell curves together. May you be forever curious and never afraid of math. I love you more than there are stars in the sky.

To my partner, Jan, who didn't divorce me (even though we're not even married). I know the whole book-writing thing sucked for you. So, thank you for living in squalor while I let the house fall down around us. And thank you for showing up as a partner to keep our lives running. I appreciate all the times you told me you were proud of me, and that you knew that I could do it. (I was skeptical—but, thankfully, you were right.) If you're reading this, that means the book-writing is over, and now you have your girlfriend back. Take me out for some cake?

To my dog, Boon, who sacrificed countless long walks and laid faithfully at my side until this book was done. You're such a good boy, and you will always be my fur baby. Woof.

Contents and Executive Summary

Foreword viii

Introduction x

Chapter 1: Conversion Design Drives Impact 1
Learn about the Conversion Design process, as well as why and how it drives business impact.

**Chapter 2: The Understand Phase:
Uncovering Impactful Insights 21**
How to find customer problems to solve, which are likely to drive business impact.

**Chapter 3: The Hypothesize Phase:
Think Clear, Logical Thoughts 43**
Think like a scientist with null hypothesis testing and by collecting highly reliable evidence.

**Chapter 4: The Prioritize Phase:
The Work and the Workflow 67**
Pick the most impactful ideas to work on and optimize your workflow to drive more impact, faster.

**Chapter 5: The Create Phase:
Set Your Idea Up for Success 91**
"Design like you're right" by crafting the best variant possible to try and beat the baseline.

Chapter 6: The Test Phase: Test Like You're Wrong   131
> Learn about good experimental design to ensure that you end up with highly reliable evidence that you can learn from.

Chapter 7: The Analyze Phase: Learn from the Data   169
> Look skeptically at the data to see if there's any evidence of impact, so that you can learn something new and make the right call.

Chapter 8: The Decide Phase: Make an Optimal Choice   203
> Avoid bias in your decision-making and deliver win-win changes that make things better—not just different.

Chapter 9: Scale It: Drive Impact Across Your Organization   235
> Align your organization around the Conversion Design process to unlock optimal decision-making at scale. Then watch your product improve and your business grow!

Final Thoughts   267

Appendix   269

Index   279

Acknowledgments   287

About the Author   292

Foreword

Lucky us! We've worked with Erin Weigel for the best part of a decade since we collaborated at Booking.com in Amsterdam. She joined us as a designer with a particular passion and enthusiasm for seeking truth and certainty in a largely uncertain world. Together, Erin and the rest of us learned through rigorous experimentation that much of what we *thought* we knew was *very wrong*, and that most of what we *didn't know* was *discoverable*, given the right environment, process, and mindset. Booking.com taught us many things, but most importantly, we learned to be humble and open to learning.

For those of you who have worked in organizations where experimentation forms a central part of the decision-making process, aspects of Erin's approach to working might feel intuitive. However, in boardrooms and shop floors around the world, most people rely on instinct to make decisions that could be improved through experimentation—and they should read *Design for Impact*.

Few people understand this problem space better than Erin. She is a living, breathing evidence-collecting machine for whom experimentation is the first tool she reaches for to solve complex problems and answer ambiguous unknowns. She shares her experiences working in tech, traditional retail, and customer service, allowing her to draw from diverse examples and highlight fascinating stories. Not only that, but she's funny as well. Look for her asides and footnotes if you want a good laugh.

Erin's methods will guide you through a way of operating that will give you superpowers. You'll discover how to conceive of, design, implement, and measure change at a pace and scale unachievable through any other means. Read about examples of experiments that had transformative impacts on their respective organizations. Most importantly, learn how you, as designer, product manager, or business executive, can leverage your experience in support of proper, measurable evidence.

Experimentation has helped us and the companies we worked for to build better products for our customers. We know this book will help you learn the tools to do the same.

—Stuart Frisby
Director of Design at Deliveroo and
former Director of Design at Booking.com

—Lukas Vermeer
Senior Director of Experimentation at Vista and
former Director of Experimentation at Booking.com

Introduction

There I was, on-stage at a conference in front of a whopping 700 people. I'd just breathlessly given a 30-minute talk about what I'd learned from running more than 1,200 experiments as a Principal Designer at Booking.com.

It was Q&A time, and the questions popped quickly onto the screen behind me. My adrenaline was sky-high, and after a moment that seemed like a lifetime—the most popular question soared to the top:

"Could you make a checklist of where to start when you improve a design?"

"A checklist, huh?" I pondered out loud. "Fitting everything into a checklist will be tricky. There's a lot to think about. In fact, it's less of a checklist and more of a book."

Shockingly, people clapped at the mention of a book. I'd had coworkers tell me to write a book before, and I thought they were just being kind. But now I had a room full of strangers egging me on about it. It was at that moment I realized that I should—at the very least—try to write a checklist.

"OK," I said. "I'll get back to you on that."

Fast-forward three years and a couple hundred pages later. You're holding my incredibly inflated, "I'll-get-back-to-you-on-that."

Anyway, this book is about the *fun*damentals of experimentation. Unlike other books about the topic, I put a lot of emphasis on the fun part. Whenever possible, I tell silly stories, find memes to capture the essence of an idea, or draw pictures. If you're a serious person who doesn't like fun, then this book is not for you. But if you're a curious person who likes to learn through play—you've come to the right place.

As a former manager of a small chain of toy stores, I learned that kids run their own experiments every day through play to learn how the world works. For whatever reason, adults get boring as they grow older. They forget to play, and as such—they forget to learn.

This book is for the young at heart who know that play through experimentation is serious, fun business. And it can also be seriously fun and effective in business.

I realize that, because experimentation covers a lot of territory, not everyone will be interested in every chapter of this book. One of my favorite sayings is, "You also gotta know what you don't need to know." So, please feel free to duck in and out of chapters as you please and only take what you find most useful. However, if you're one of those weirdos (like me) who likes to learn a lot about a lot—and that's why you got into experimentation—then you'll get the most out of this book by reading each chapter in order.

This leads me to some small disclaimers: This book is not a textbook, and I am not a statistician. This is a trade book, and although I'm not a math person, I do have a lot of "mathy" friends who've helped me write and review it. That said, this book's purpose is to give non-sciencey, non-math folks (like me and maybe you, too) some useful tools and OK-ish enough understanding to experiment better in a business setting.

This point is: I hope this book piques your curiosity and that you find it useful. And if you don't—at least I put a checklist at the end!

CONVERSION DESIGN PROCESS

CHAPTER 1

Conversion Design Drives Impact

THE POINT

Many product design and development teams use opinions to make decisions. A better approach is to gather reliable evidence, measure the impact of their work, and make optimal decisions by learning from the Conversion Design process.

Have you ever sat around a table at work and discussed endlessly which version of a design is best? I know I have. Early in my first real tech job at a meal delivery service, I was asked to redesign part of a home page. The goal was to sell more of our secondary product line, which was meal replacement shakes. With the task in hand, I got started.

I designed one option with photos of people. Product photos dominated a second design, and a third one used only icons and words. I even had a fourth version with icons, people, and products, which, admittedly, was a bit much. Then our team had to decide which version to use (Figure 1.1).

FIGURE 1.1 Four different versions of the design execution to choose from.

We each shared our opinions:

- The marketer wanted pictures of people to "create an emotional connection."
- The business manager wanted product photos because it "was clear what we were selling."
- The developer wanted only text because it would "be quick to do."
- And, I, the designer, wanted icons and text because it "looked less cluttered."

These were all reasonable options, but which one would have the impact we were aiming for? We needed to break free from our opinions to focus on user needs and business goals. To do that, we went outside the office and did some quick preference tests with passersby.

"Which one do you like best?" we asked. After some feedback from strangers, we made a decision and rolled out the new design.

My question at the end of all this was: "Did the new design have the impact we wanted? Was it going to sell more shakes?" At the time, there was no way to know. The results were as "shaky" as the product we were trying to sell.

Even if we had looked at the sales data before and after the change, we wouldn't have been confident that the change we made caused any of the impact that we saw.[1] That's because the change wasn't isolated. The data could have been influenced by something else, such as the time of year, a special offer, an ad, etc. In this scenario, the data was iffy at best. It could have given us a clue, but it couldn't give us *confidence*. In the end, we made an opinionated best guess. We released the "nicest looking" option with just logos and some text.

Conversion Design Is Evidence Collection

Conversion Design is a way to work that moves teams *beyond* opinion. It uses research, A/B testing, and data analysis to create reliable evidence, which enables consistently better decision-making. With this process, you can know with confidence that changes are resulting in the intended impact. In other words, did the change you made actually help you achieve your goal?

Evidence is at the heart of Conversion Design. It draws on the experiences of three disciplines (Figure 1.2):

- **Design** with its focus on solving customer problems
- **Science** by way of unbiased, evidence-based decisions through the use of the scientific method
- **Business** from a value-creation and transfer perspective

---

[1] This type of experiment is called an *interrupted time series*. It's unfortunately not very reliable unless it's repeated multiple times.

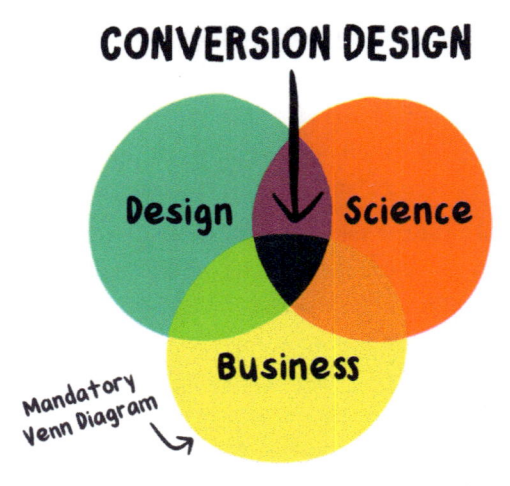

FIGURE 1.2 Conversion Design combines design, business, and science.

Accurate Measurement Beats Opinion

A core model from science that's used in Conversion Design is the *hierarchy of evidence*.[2] This is a way to assess the reliability of different kinds of data or information gathered from studying something (Figure 1.3).

FIGURE 1.3 The hierarchy of evidence as used in evidence-based medicine.

2. Patricia B. Burns, Rod J. Rohrich, and Kevin C. Chung, "The Levels of Evidence and Their Role in Evidence-Based Medicine," *Plastic and Reconstructive Surgery*, 128, no. 1 (2011): 305–310, www.ncbi.nlm.nih.gov/pmc/articles/PMC3124652/

The hierarchy of evidence's broad base represents expert opinion. Although it's the least reliable form of evidence—because of the likelihood that it's biased—it's nonetheless a decent starting point. True experts are people who have learned from a lot of evidence further up in the hierarchy.

> **TIP** TRUE EXPERTS OFTEN DON'T CALL THEMSELVES EXPERTS
>
> A relevant quote attributed to Albert Einstein is: "The more I learn, the more I realize how much I don't know." Beware of people who claim unwavering expertise in a topic but who don't have strong evidence to back up their claims. Seek out experts who want their opinions tested. True experts often doubt their own knowledge and embrace the fact they don't know everything. Self-proclaimed "experts" are likely to state their opinions as facts and just push their opinions through. ■

One step up from expert opinions are *observational studies*. These include usability tests, shadowing sessions, and surveys. A drawback of observational studies, as with expert opinion, is the possible bias of the observer.

The next step up is *randomized controlled experiments*. A randomized controlled experiment is a scientific study in which participants are randomly placed into different groups. *Random* means that each visitor has an equal chance of being assigned to different groups. Splitting participants into randomized groups is a simple way to collect reliable evidence because it allows you to "isolate" a variable you want to learn about. Isolation ensures that the variable you're examining is the only thing that's different among the groups. This helps you know that if the results are different, it's likely that your isolated variable is what made the difference.

Well-designed randomized controlled experiments provide very reliable evidence because they enable relatively accurate measurements. As the pioneering American computer scientist Admiral Grace Hopper once said, "One accurate measurement is worth a thousand expert opinions."[3]

Finally, at the top you have a *systematic review*. This occurs when scientists look critically and without bias at all the evidence collected about a given topic to form a well-rounded conclusion. This review provides a holistic summary of the findings from all available knowledge.

---

3. MacTutor, "Quotations: Grace Hopper," School of Mathematics and Statistics, University of St Andrews, Scotland, https://mathshistory.st-andrews.ac.uk/Biographies/Hopper/quotations/

A/B Testing Your Way to Reliable Data

A/B testing is a common form of randomized controlled experiment. To run an A/B test, you make a change to one part of a user's experience. Then you expose a random half of the users to the original experience and the other random half to the changed experience. To understand how it works, think back to the story of my team making a change to a home page at the beginning of this chapter. Figure 1.4 illustrates the original and the change.

FIGURE 1.4 The parts of the home pages that were tested are highlighted.

The original is called the *control*. It's the "A" in an "A/B test." You might also hear some people call it *base*, which is short for *baseline*. The change is called the *treatment* or *variant*. It's the "B" in an "A/B test." Instead of showing the change to every home page visitor, it's only shown to a random segment of the visitors. The rest of the visitors see the original version. After the two versions (Figure 1.5) are shown to enough visitors, you can stop the experiment and learn if there's an impact on sales from the result.[4]

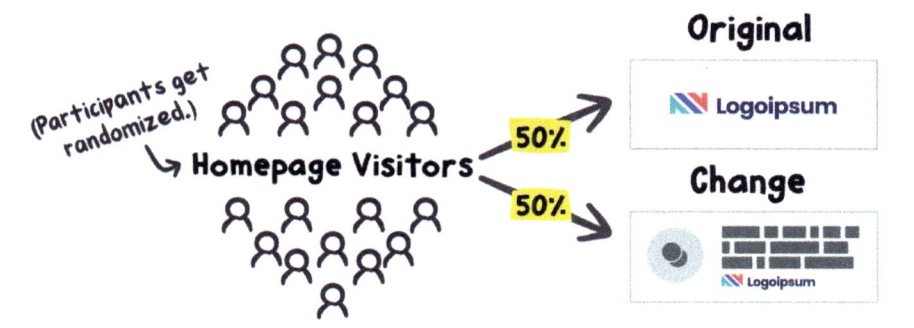

FIGURE 1.5 Here, 50% of visitors see the original (base) and the other 50% see the new version (variant).

What do you think the results will be?

- Will the change be too small to detect—and have an "inconclusive" result?
- Will it do better than the original—and have a "positive" result?
- Or, will the original do better than the change—and have a "negative" result?

Changes often make no measurable difference. But sometimes you get surprising results. In this example in Figure 1.6, the change performed 3% *worse* on overall sales than the control.

This change unintentionally made the customer experience and business outcome worse. Because the sample was randomized, we knew that the decrease in sales was more likely due to the change in the variant rather than chance. This effect was something we couldn't have known confidently without an experiment. You need to know the impact of your work because your job is to create value for customers and the businesses that serve them—*not* destroy it.

---

4. "Enough" is a tricky concept. It's so tricky, in fact, that much of Chapter 6 is dedicated to it! Stay tuned for more details on "enough."

FIGURE 1.6 In this case, the variant performed worse on the success metric. Sales of the secondary product line, meal replacement shakes, were less with the change.

Another Way to Think

Unlike other design processes, which are based on linear thinking, Conversion Design is based on systems thinking. Systems thinking, as illustrated in Figure 1.7, is a practice and set of tools that helps you understand and capture things in a dynamic and interconnected way.

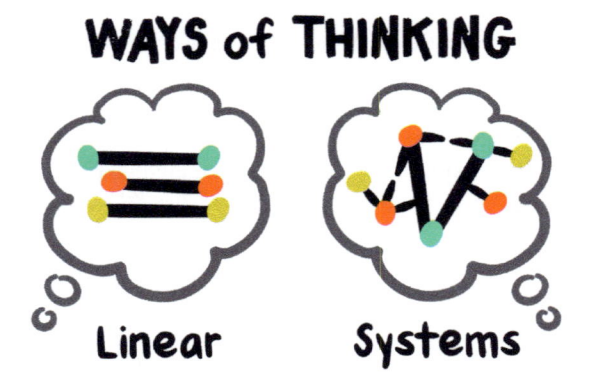

FIGURE 1.7 Linear thinking (left) leans on simple cause-and-effect relationships. Systems thinking (right) identifies many interacting variables and how they relate to one another.

Systems thinking lets you understand how and where multiple things interact, and ultimately—their impact on one another. With it, you can represent and understand the world around you more accurately to make optimal decisions in the face of complexity.

Conversion Design works much like a machine (Figure 1.8). Teams work together and lean on each other's strengths at different points in the Conversion Design process by representing a wide range of disciplines. Everyone plays an important role to get through each step.

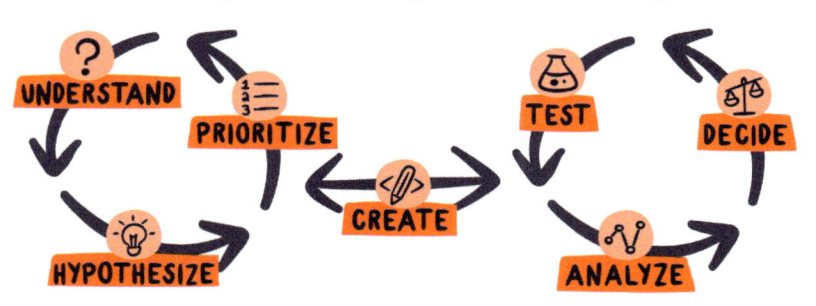

FIGURE 1.8 The Conversion Design process enables teams to learn from reliable evidence to help them make better decisions.

As the gears in a machine turn and flow into and out of one another, they create an "output." This output creates a loop (a "flywheel") that amplifies the whole system by reinfusing energy back into the process (Figure 1.9).

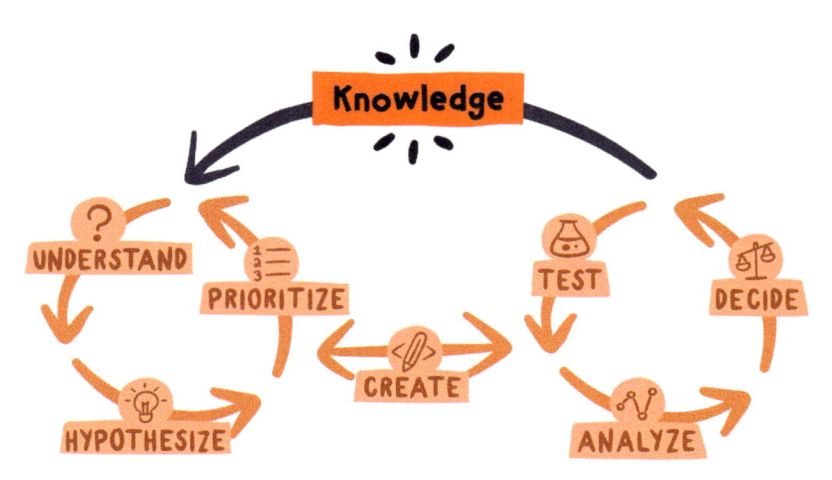

FIGURE 1.9 The "flywheel" throws off energy in the form of collective knowledge, which gets reinfused back into the system to make it stronger.

In the case of Conversion Design, the reinfused energy comes in the form of collective knowledge—the combined brainpower of a group of people. As you gain knowledge, you take it with you as you begin the process again. The main learning happens during the testing phase of the Conversion Design process in which a cause-and-effect relationship is examined. This cause/effect relationship is a hallmark of linear thinking. In this way, Conversion Design benefits from both ways of thinking: systems thinking to embrace the complexities of the real world and linear thinking to isolate the cause/effect relationships that teams can learn from.

Conversion Design Creates Value

Value creation—and the eventual transfer of that value to the next-in-line stakeholder—is the heart of all work. Conversion Design helps teams identify and deliver value, as well as finding and avoiding damage to value that already exists.

For customers, value comes in many forms. Common ones are time saved, money saved, convenience, and social connection. Value for business is often measured by growth in company profits through an increase in customer purchases, as well as increased customer attention and action. For employees, value might be compensation, benefits, and (hopefully) a sense of purpose.

Just as Conversion Design creates knowledge as an output, it creates customer value as well. It's represented as a dashed line to show it's not guaranteed, it's intermittent (Figure 1.10).

Unfortunately, many people conflate the idea of "conversion" with just being the sales part of business value creation. In Conversion Design, *conversion* stems from the Latin word, *convertere*, which means *to change*.[5] The root of conversion has nothing to do with money, value, or sales. It's about transformation.

---

5. Online Etymology Dictionary, "Convert (v)," www.etymonline.com/word/convert

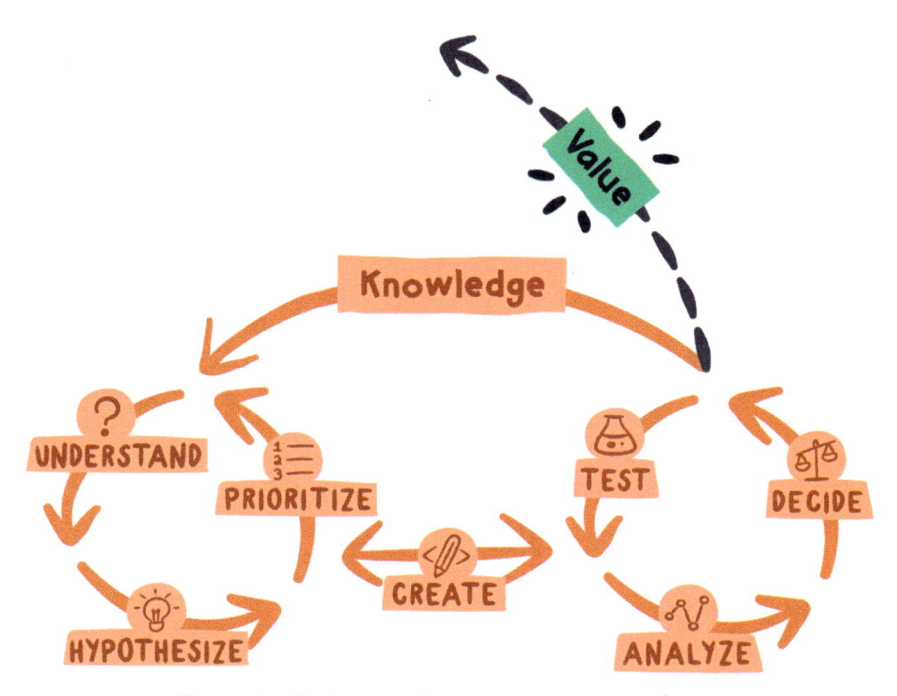

FIGURE 1.10 Conversion Design sometimes creates customer value as an output, too. But that output gets infused into another system.

The word *design* is also often misunderstood. It's typically mistaken for "making something look nice." But at its core, design gives shape to ideas and makes them real. In other words, "design is the rendering of intent," as Jared Spool, prominent designer and founder of the design school Center Centre, calls it.[6]

"Conversion," when combined with "design," means "to create intentional change." But not just any change: improvement. True product improvement creates value for customers. And though Conversion Design might sound fancy, it simply means good design that fulfills its purpose of creating value for customers and the businesses that serve them.

---

6. Jared M. Spool, "Design Is the Rendering of Intent," https://articles.centercentre.com/design_rendering_intent/

The Value Cycle

Eventually, value created by the Conversion Design process for customers gets transferred back into the business. This concept is captured by the "value cycle," which is a key concept within Conversion Design, as shown in Figure 1.11.

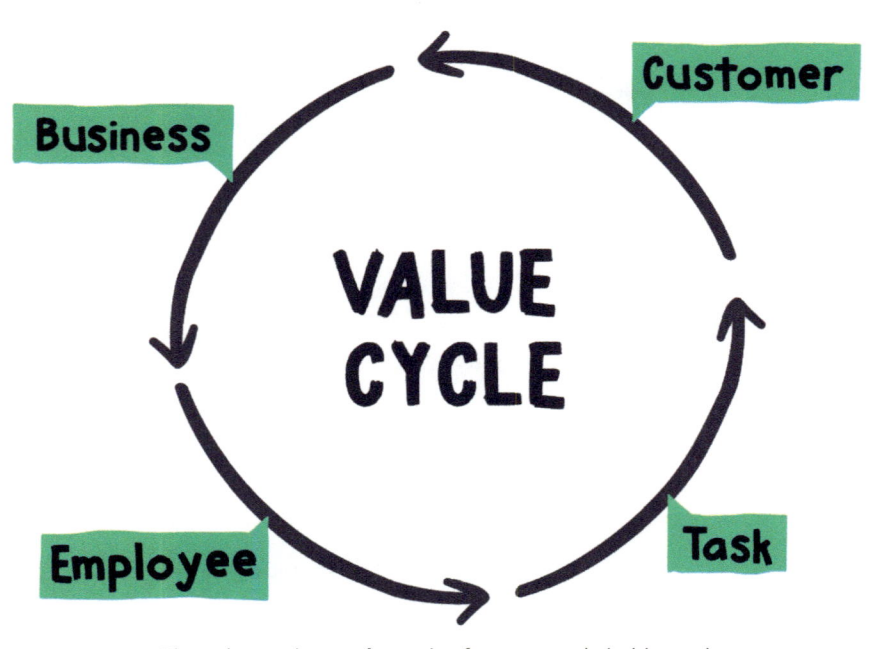

FIGURE 1.11 The value cycle transfers value from one stakeholder to the next.

The value cycle is a loop that represents how a company, such as a product or service, creates and delivers value to customers.[7] Because it's a loop, companies eventually get some form of that value back. The company uses this returned value to continuously improve or create new products through its employees and the tasks they do, which starts the cycle again.

---

7. Peter Cohan, "Riding the Value Cycle," *Business Strategy Review*, London School of Business, Summer 2008, www.london.edu/-/media/files/publications/bsr/riding-the-value-cycle.pdf?la=en

The goal of Conversion Design is to feed this value cycle by understanding what changes (if any) are needed, who those changes would benefit (or not), and most importantly, ensure that any changes made lead to the intended effect. In a business's value cycle, Conversion Design begins when the business invests in the Conversion Design process as the employee's default way of working. When you put it all together, Figure 1.12 shows how the Conversion Design process "shakes out" graphically within a business value cycle:

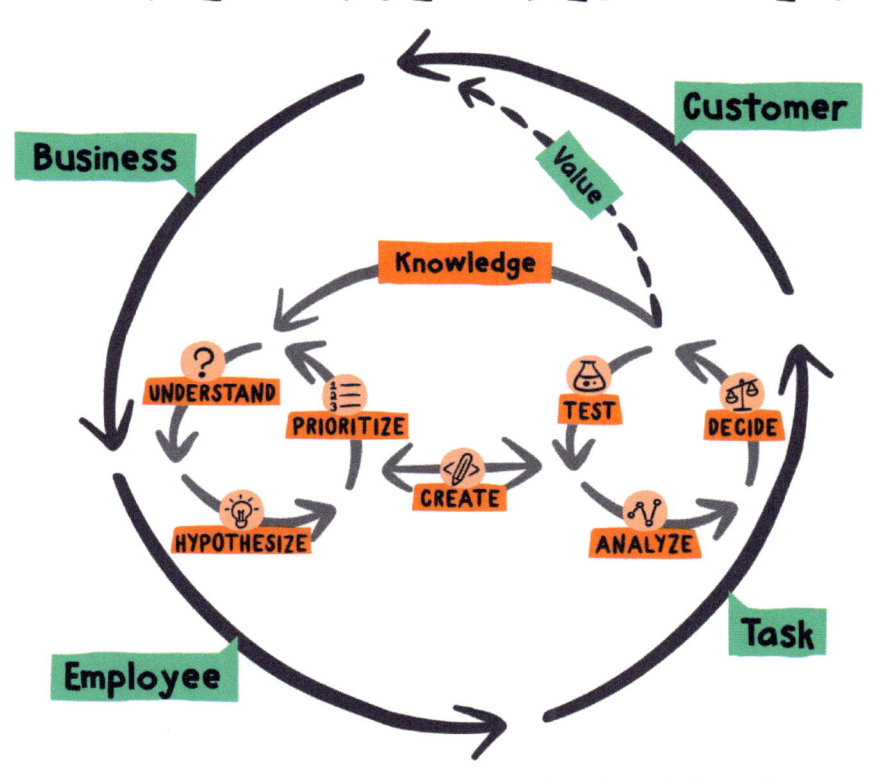

FIGURE 1.12 The Conversion Design process drives the value cycle forward.

EXPERT ADVICE FROM THE FIELD

DIVERGE AND STAY OPEN WITH KEVIN ANDERSON

Many designers use the linear design thinking process called the *Double Diamond* to guide their work.[8] It's a two-phase process, as shown in Figure 1.13.

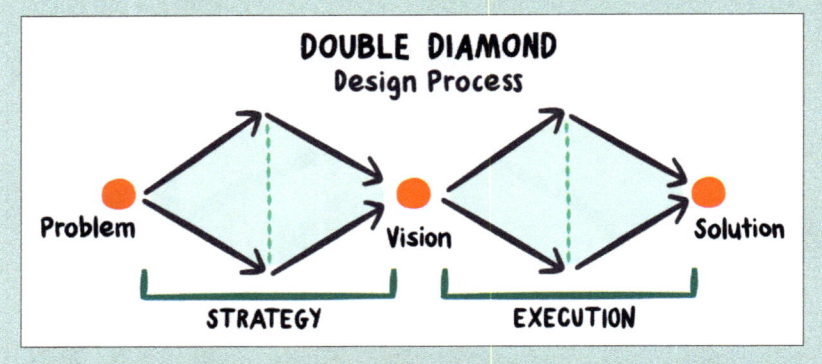

FIGURE 1.13 The Double Diamond design process diverges and converges to end up with a single solution.

8. Design Council, "Double Diamond: A Universally Accepted Depiction of the Design Process," www.designcouncil.org.uk/our-resources/the-double-diamond/

In the strategy phase, you "diverge"—go conceptually broad—to learn about the problem and generate many ideas to solve it. Then you "converge"—narrow down your options—to decide what solution to execute. In the execution phase, you "diverge" by exploring the different shapes the solution could take. Then you "converge" to decide which specific execution of the solution you'll deliver.

In my experience, designers often converge quickly and in isolation. When they do this, lots of great ideas never see the light of day. And in experimentation, the more ideas, the better. What I'd love to see is designers be more open with their ideas—share them raw, early, and often to get input. I believe that designers don't do this because teammates and stakeholders don't know the best ways to give feedback. Asking for and giving feedback are skills that teams should work to improve. My advice is that if you use the Double Diamond as a guide: diverge and stay open. Then share—and test—all the possibilities.

Kevin Anderson
Product Manager of Experimentation at Vista

Kevin is an active member of the experimentation community and founded www.experimentationjobs.com *in 2020 to help practitioners find great places to work.*

Conversion Design Leverages the Compound Effect

Many companies follow outdated product development processes, which don't include systems thinking, the scientific method, and experimentation. They often release products without understanding the most likely impact of their changes. This approach results in releasing a mix of positive, negative, and inconsequential decisions. The positive and negative decisions cancel one another out, so the results reflect the mixed success rate, shown in Figure 1.14.

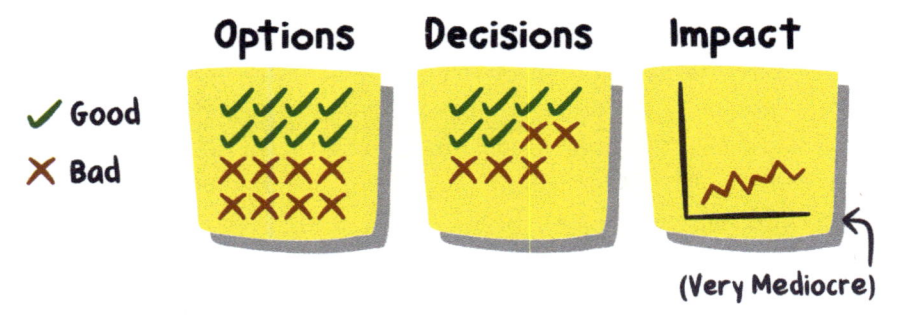

FIGURE 1.14 Typical product development results in both good and bad decisions, which leads to varied impact.

With Conversion Design, teams learn and improve. As a result, the products that teams build improve, too. The Compound Effect is when many small, good decisions combine over time to create an exponential effect.[9] Conversion Design creates this effect by lowering the rate at which teams make bad decisions. To get the most benefit from the Compound Effect, you should only keep the *positive* changes. Toss the unimpactful and negative changes in the trash. In time, good builds upon good, upon good.

The results of those consistently good decisions executed over the course of a long period of time could look something like Figure 1.15, the coveted "hockey stick" growth line that most businesses aim for.

---

9. Darren Hardy, *The Compound Effect: Jumpstart Your Income, Your Life, Your Success* (New York, NY: Vanguard Press, 2010).

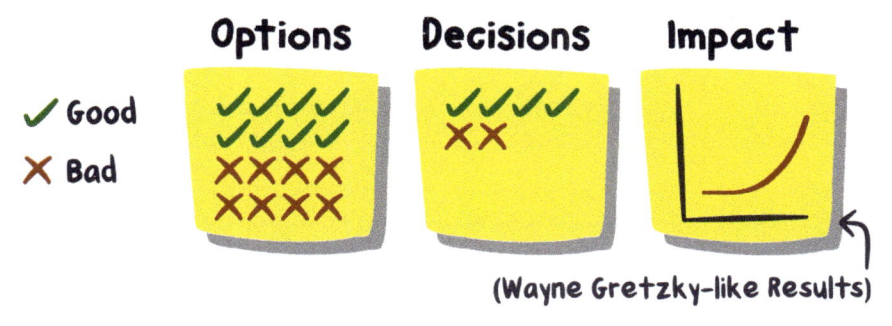

FIGURE 1.15 Many good decisions compound over time create an exponentially positive effect.

Initially, the line stays relatively flat and grows slowly. This gradual incline is normal at first. Eventually, when the Compound Effect takes hold, it spikes quickly. Working this way helps you create and receive value faster than your competitors, which gives you a big advantage. When you stay agile and learn fast, you can adapt to the changing competitive landscape and focus on delivering the most impactful work.

Good Process Is Important

At this point you may think, "Another process?! That's exactly what we *don't* need." And that's understandable. Processes get a bad rap because businesses pile one on top of another, which can slow things down. But the right type and amount of process can ensure quality, speed, and alignment. When you don't have a clear process, it leaves room for stakeholders to disagree on the approach. This can cause different—sometimes conflicting—decisions. And when people don't have the same goal and the same plan to get there, team confidence and employee engagement drops.

Tried-and-true processes overcome this. When people know what needs to be done and how, they can understand where they add value within the process. Better yet, they understand where their teammates bring value, too. This understanding of one another's strengths cultivates healthy team dynamics, which (in addition to helping you not hate your job) increases the team's collective intelligence. Collective intelligence is the high-quality thinking and creativity that emerges within a group. It's powerful because when diverse minds work together, two benefits emerge:

- Lots of unique perspectives, which build empathy. And expanded empathy leads to more effective ideas to serve your ideal customers.
- A broader skill set, which increases the pool of skills available, and leads to better execution of those ideas.

When you combine these benefits, your team can outperform any narrow-minded competition. But you need a solid and inclusive process to get everyone on the same page.

The Seven Steps of Conversion Design

At the end of the Conversion Design process, you should know with confidence that the change you tested was good, bad, or made no difference. But exactly how do you get to that point? Conversion Design involves these seven interconnected steps or phases:

1. **Understand.** In this phase, you look at customer problems through the lens of business goals. The purpose is to develop actionable insights that will enable you to make measurable product improvements that positively impact both customers and the business.
2. **Hypothesize.** Next is turning those insights into testable ideas. These are thoughts or questions that experimentation can help you answer. This is done by way of the scientific method, which is a structured thought process that allows you to collect and learn from reliable evidence.
3. **Prioritize.** From your list of testable ideas, you strategically guess which ones are likely to solve customer problems in a way that positively impacts the business. Then you weigh short-term gains against long-term business goal alignment and put them in an order that makes sense.
4. **Create.** Here is when your idea begins to take shape. You design and develop a testable product change (i.e., a new feature, content refinement, different process flow, etc.) to expose to your customers.
5. **Test.** This phase introduces your testable idea to the real world via an A/B test. This part of the process is where you can discover cause-and-effect relationships that you can learn from.
6. **Analyze.** Once the test is complete, you look at the data to understand the impact of the change. At this phase, you learn if the overall effect of the change made the impact you were aiming for—or not.

7. **Decide.** Finally, you use what you learned during the process to make an optimal decision for all stakeholders.

Now let's step into the Conversion Design process and cycle through the stages. First up: The Understand phase in Chapter 2.

The Important Bits

- Teams no longer have to rely on opinion to make design decisions. They can use the Conversion Design process to collect reliable evidence to make optimal decisions.
- Conversion Design is a product development and design process, which is based on systems thinking. It captures the dynamic, interconnected nature of reality.
- Conversion Design combines three disciplines: science, business, and design, in order to drive measurable, positive impact on both the customer experience and business results by making optimal, evidenced-based decisions.
- Teams methodically work their way up the hierarchy of evidence to get reliable information to make optimal decisions.
- The most reliable form of evidence is a randomized controlled experiment. A/B tests are the most basic form of a randomized controlled experiment, and it's an important part of the Conversion Design process.
- Teams should work together and lean on one another's strengths to harness the collective intelligence of the group.
- The outputs of the Conversion Design process are collective knowledge and customer value. Value creation within a business's value cycle is what moves the business forward.
- Eventually, through experimentation and optimal decision-making, the Compound Effect kicks in. This occurs when many small, good decisions come together over time (while throwing away the bad) to make a big positive impact.
- The seven steps in the Conversion Design process are: Understand, Hypothesize, Prioritize, Create, Test, Analyze, and Decide.

CONVERSION DESIGN PROCESS

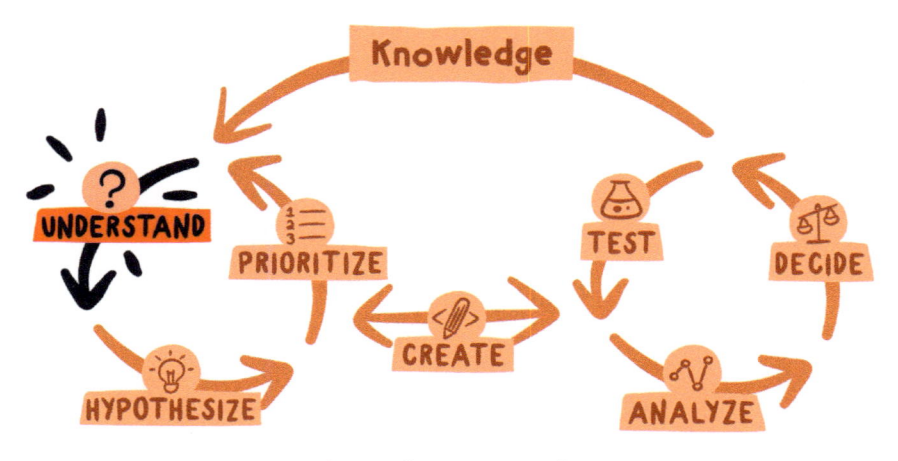

Knowledge

UNDERSTAND • PRIORITIZE • CREATE • TEST • DECIDE • ANALYZE • HYPOTHESIZE

"Research is formalized curiosity. It is poking and prying with a purpose."

— Zora Neale Hurston

CHAPTER 2

The Understand Phase: Uncovering Impactful Insights

THE POINT
Find customer problems that when solved are likely to create business value.

When I was 16, I waited tables at a local restaurant. During my first few shifts, I "shadowed" a seasoned server to learn the job. Little did I know, this training was my introduction to the world of customer research. As I shadowed, I watched my trainer, June, prepare the table, greet the guests, and take the orders. Most importantly though, I learned what diners found useful based on tip size. Tipping is a standard practice in the U.S., and typically, if the service is bad, people tip 10%. If the service is standard—it's 15%. If you do amazing, people might tip up to 18%–20%, or more!

It was real-time feedback that made me a better server. This job also taught me to identify key moments in the dining experience, such as the "two-bite/two-minute check back." That's when servers stop by the table after food is delivered to make sure that everything is OK. If the diner hasn't taken a bite after two minutes, maybe they need ketchup? Or a sharper knife? Or, perhaps after starting to eat, they could use more napkins? Or another drink? If so—I'd be right there to help.

Out on the restaurant floor, I read both the body language and the situation to understand what people were thinking or feeling. But serving customers in a digital space removes that meaningful human interaction. It makes giving great service much harder. Digital product development requires creativity to understand customer needs. However, what you lose in human interaction, you gain in rich data. This data sticks around, even when employees leave, so it builds up over time. Instead of face-to-face contact, the digital world gives you qualitative and quantitative research to get closer to the people you serve.

Research Is Formalized Curiosity

Curiosity is an important characteristic in experimentation, and research is an expression of that curiosity. As Zora Neale Hurston, an American author and anthropologist once said, "Research is formalized curiosity. It is poking and prying with a purpose."[1] Poking and prying alone doesn't make an impact, though. You must harness and direct it. The Understand phase of Conversion Design does just that by discovering real customer problems that, when solved, are likely to make a positive impact. To do that, you must focus on the top customer issues by looking at them through the lens of important business goals (Figure 2.1).

---

1. Zora Neale Hurston, *Dust Tracks on a Road* (New York: Harper Perennial, 1996), 143.

RESEARCH STRATEGY

- Actionable Insight
- Research Method
- Research Question
- Customer Top Tasks
- Business Goals

FIGURE 2.1 Research should be viewed through the lenses of your business goals.

Many companies have priorities, success metrics, and goals. If your company doesn't, you could learn what's important by understanding how the company serves customers, makes its money, and where those two things overlap. Then you can create a list of top customer tasks that align with those goals. Top customer tasks are the main things that people want to do when they use a product or service. If you need more information, use the product yourself, or watch someone else use it. You want to learn the flow and the tasks that need to be done. As you use the product, write down your observations. Questions to ask yourself are:

- Where do people make errors?
- Is any language unclear?
- Where might people hesitate?
- What common mental models and design patterns are broken?
- Does anything seem particularly slow?

Once you have a list of observations, it's time to define your research question.

Ask Questions to Collect Data

A *research question* is a clear, simple question that you want to answer by studying and collecting information. Research questions can provide answers in two forms of data: qualitative and quantitative. Qualitative data are facts that don't involve numbers. Some examples of qualitative research are behavioral observation (ethnography), open questions on surveys, and diary studies. Quantitative data are facts expressed as numbers. Some examples of quantitative research are analytics, A/B testing, and closed questions on surveys. Although many types of research questions exist, the following ones provide preliminary and secondary insights, which give you a solid foundation for research within experimentation. Two question types to uncover preliminary insights are the following:

- **Exploratory.** These questions are broad and help you observe and understand something without bias. Use them when you want to learn broadly about something where there's little existing knowledge or understanding. An example is: "How do people decide what refrigerator to buy?"
- **Descriptive.** These questions seek to explain when, where, or how many times something happened. They are the most basic kind of research question. An example might be: "How many people clicked on the main call-to-action button within the last month?" It's simple and clear.

After you find something interesting to dig into, then ask the "why" and "what-if" questions:

- **Explanatory.** These questions aim to explain *why* something happens. They are often used to delve deeper into a preliminary insight. They usually start with "why." An example is: "Why do people drop off in the payment step of the buying process?"
- **Relationship-based.** These questions seek to understand if one thing impacts another. This question type provides the backbone of experimentation. An example might be: "Will more people make a purchase if we add a product comparison tool?"

After you write your research question (what you hope to learn), then you choose your research method (what you'll do to learn it).

Choose the Right Research Method

Different research methods—just like research questions—uncover different types of insights. Christian Rohrer, a design strategist and author, created the "Landscape of User Research Methods" (Figure 2.2) to help you choose your research method.[2]

---

2. Christian Rohrer, "When to Use Which User-Experience Research Methods," Nielsen Norman Group, July 17, 2022, www.nngroup.com/articles/which-ux-research-methods/

LANDSCAPE of USER RESEARCH METHODS

TYPE OF USE: ● Natural ■ Guided ▲ Decontextualized

BEHAVIORAL
What People Do

- ■ Usability Tests
- ● Shadowing Sessions
- ● A/B Tests & Data Analytics

ATTITUDINAL
What People Say

- ● Customer Feedback
- ▲ Interviews & Focus Groups
- ▲ Surveys

QUALITATIVE
Why & How to Fix

QUANTITATIVE
How Much & How Many

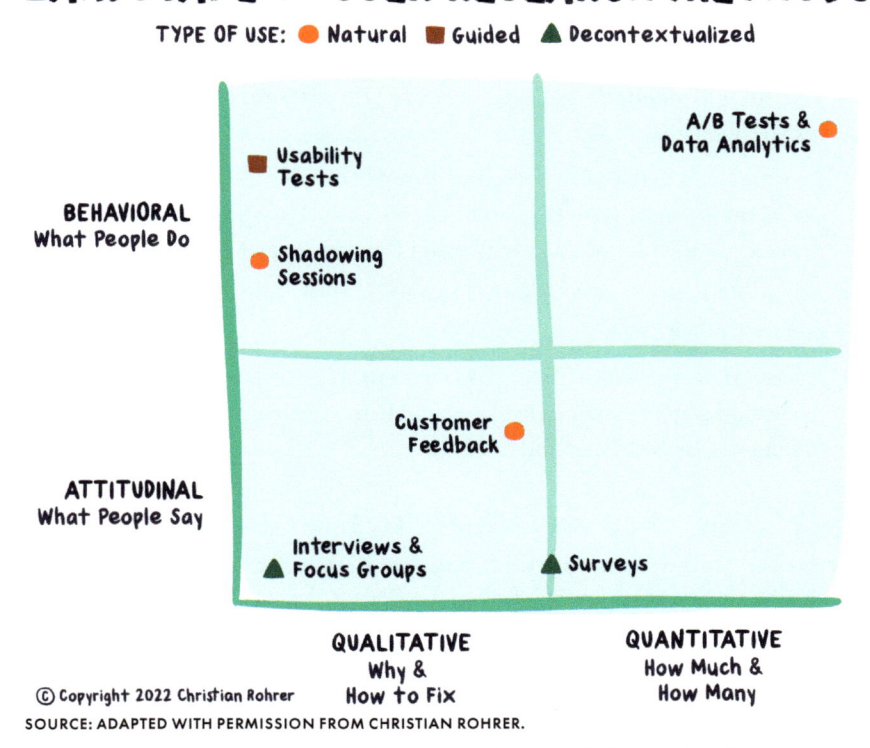

© Copyright 2022 Christian Rohrer
SOURCE: ADAPTED WITH PERMISSION FROM CHRISTIAN ROHRER.

FIGURE 2.2 This matrix captures the spectrum between behavioral and attitudinal insights and plots them against qualitative and quantitative data output.

One axis describes what the research actually documents—for example, behavior or attitude. The other shows if the result gives you qualitative or quantitative information. Research methods dot the landscape, and their position shows the types of insight they uncover: qualitative, quantitative, or a mixture of both. The landscape also illustrates the three types of product use within research:

- **Natural use.** People do what they want without influence.
- **Guided use.** People get tasks to complete and prompts to follow.
- **Decontextualized use.** People are asked questions about a product, but not while using it.

Consider these things when you choose a research method:

- **The research question.** Start your study with a clearly defined research question that ties back to a business goal.
- **The insight needed.** The type of insight depends on whether the research must identify needs or inspire ideas about how to address them.
- **Product maturity.** If the product is immature, it's likely to need discovery. If the product is mature and insights exist, then understanding how to fix known issues is more important than uncovering new ones.
- **Available resources.** Timelines, employee skills, and research investment vary from business to business.
- **Access to data and information.** Data protection laws and security mean that not all information is available to everyone. Learn what type of data can be collected and used.

> **TIP** WHEN TO USE QUANTITATIVE VS. QUALITATIVE
> If no clear user problem or need exists, quantitative research helps you dig into the "what, how many, or how much." If a clear need exists, then qualitative research helps address the "why" and "how." ∎

Choose the Right Research Environment

The last thing to consider is your research environment. Environments shape human behavior, so your choice influences the quality of your observations. Controlled environments are professional spaces where researchers can conduct the test and prepare logistics beforehand. Examples are the researcher's office, usability labs, and observation studios. These are good for in-depth general insights, such as usability baselines, moderated studies, or user interviews.

Remote environments are software solutions that allow participants to choose a spot that works best for them. Examples are remote usability study software, card-sorting tools, and digital diary studies. These are good for lightweight, general feedback on how people interpret content and find basic usability issues.

Natural environments are where your participant typically uses your product or service. Examples are the participant's home, school, or workplace.

Natural environments are important for products that involve human, physical object, or location-based interactions. If context is crucial, go to where the issues happen to observe exactly what's going on.

> **TIP** "BYOD" (BRING YOUR OWN DEVICE)
>
> Regardless of where you do your research, be sure to have participants use their own mobile device, computer, or hardware. When participants use an operating system or device they're not used to (or familiar with), it's hard to tell if an issue is caused by the product you're testing or the hardware and its software. Observing a participant using their own device also uncovers information about these customer features:
>
> - **Settings**, such as device language, notifications, sound and haptics, special keyboards, brightness and color, etc.
> - **Shortcuts**, such as prioritized apps in the toolbar, gestures to activate certain features, navigation between screens and apps, autocomplete, etc.
> - **Adaptations**, such as personal changes or hacks made to the device, so it works how they want it to. ∎

Explore Data from Behind Your Desk

Data can be collected continuously and analyzed whenever a question arises. Many ways to get inspiration for your experiments are not only fast and affordable, but they can be fun, too. A great place to look first, which many companies collect in the background, are in the reporting analytics. These are numerical data about a product's performance and usage such as Google Analytics. Here are some research questions to get you started:

- **Where do people drop off?** Why do they leave?
- **Where do they spend a lot of time?** Does it make sense that they spend a lot of time there?
- **Where do they go when they leave an important screen?** Do they go where they're supposed to go? Or do they get lost somewhere?

The answers to the bolded descriptive questions can lead you toward some deeper qualitative research via the follow-up explanatory or interpretive questions. Click maps are a visualized form of reporting analytics. They show (Figure 2.3) where and how often people interact with elements of a user interface (UI).

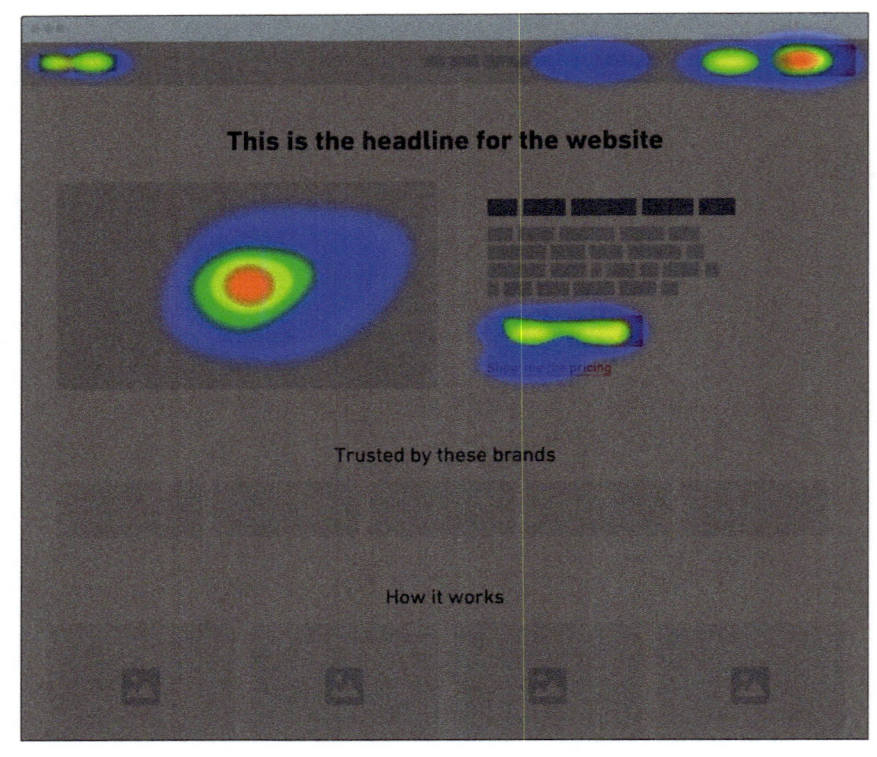

FIGURE 2.3 This click map shows where people click on a screen. You can understand a visitor's intent by looking at what people click on most.

All clicks, even ones on noninteractive elements, are logged. If you notice many people clicking on noninteractive elements, ask yourself: "Why do people click there?" Then try these follow-up research questions:

- **Did the user want more information?** If so, what information?
- **Did they think it was a link?** If so, where did they hope it would take them?
- **Was it a misclick?** Were they trying to click something else? If so, what?

> **TIP** TEST REMOVING THINGS
>
> Highly impactful products pack every pixel with value. If you notice that people don't read or interact with something, and it's not legally required—test removing it. Removing things can be just as good as—and sometimes better than—adding or changing things. ■

Customer Service Contact Logs

Customer service agents deal with all the problematic issues that a product creates or doesn't solve. Most customer service systems allow representatives to tag issues with specific topics. Topic tag examples might be:

- Account access issues
- Subscription cancellations
- Failed payments
- Billing questions

> **TIP** CREATE A TOP-10 LIST OF CUSTOMER PROBLEMS
>
> From your tagged customer service data, create a top-10 customer issues list. Note how often the issues happen, as well as time-to-resolution. With this information, you can quantify the impact of the issues for both the customers and the business. Time wasted for customers and money spent on the employee who handles each issue make great metrics to track and measure. Share the top-10 list through the company, and eventually the issues on the list will change as problems get solved through experimentation. ■

Review, Rating, and Reputation Sites

Two types of users, promoters and detractors, can provide a wealth of insight. Promoters are people who enjoy and proactively discuss a product or brand in a positive way. They are a company's most valuable customers because their opinions often influence the opinions and actions of others. Detractors are people who have a mediocre or bad experience and actively talk poorly about your company or product. Common places to find your promoters and detractors are reviews, ratings, and reputation sites (Figure 2.4).

These tools, which are run by third parties, provide a way to understand people's real experiences with and feelings about a product or business. Dig into both the combined ratings (aka aggregate score) and the individual reviews and ratings. If the ratings and reviews look good, awesome! This reassures potential customers. If they aren't, dig into the "why." Look for negative patterns in the reviews and experiment with them.

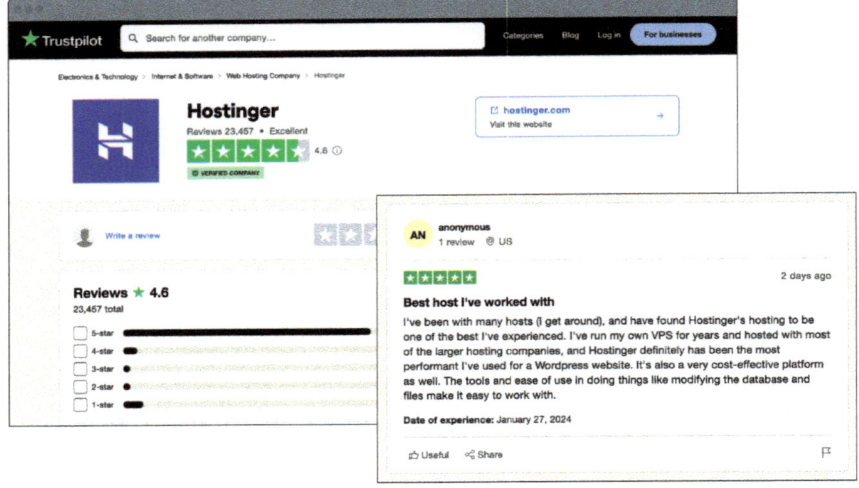

SOURCE: TRUSTPILOT.COM

FIGURE 2.4 Rating and review sites show the combined rating and individual reviews of all customer responses.

> **TIP** REDDIT, SUBSTACK, AND YOUTUBE CAN BE RESEARCH GOLDMINES
>
> If your product is discussed online in forums such as Reddit or Substack—read and act on the feedback! These folks are often passionately for or against your product, and they aren't shy about sharing their honest opinions. ∎

Surveys

Micro surveys are one to two questions that are targeted to appear at a specific time when someone uses a product. They can help you quickly understand what your users think, do, and feel. Because they're short, they take little time to design, run, and analyze. Depending on the design, they can collect qualitative data, quantitative, or both. The type of data you get depends on the types of questions you ask—open or closed. Open questions allow people to answer in their own words in a text field. This data gives people freedom to express their thoughts with nuance, which gives you a rich form of qualitative data. Closed questions provide predefined options for respondents to choose from, such as yes/no, multiple-choice, or ticked box options. This type of survey question gives you quantitative data.

Experiment to Learn

Another way to understand the dynamics of a product is to "play" with it. Ask yourself these questions to get ideas about what to "play" with:

- **What information do customers need?** Is that information where they expect it? Can everyone access it? And does it load fast?
- **What do the headlines and subheads say?** Do they contain the words the customers use and are looking for? Do they make sense?
- **What are the most-used actions?** Can everyone access them? Are they easy to do?

Answering these questions in an unbiased way is hard—especially if you're very familiar with your product. You can use A/B tests to move content, remove elements, and emphasize things you have evidence to believe are important so you can actually learn if they're important or not. This may seem irresponsible, but "purposeful play" is an effective and fun form of exploratory research that can lead to big customer and business impact.

Experimenting to learn helps you understand the value of each element on the page in its current state. Once you understand how it is or is not working, you can design meaningful improvements. Redesigns often fail because of the number of assumptions that get made about what's important. But big design leaps are often easier after experimenting to learn. It's possible that moving, changing, or removing one thing can make a whole redesign fail. Experimenting to learn helps you prioritize impactful content and interactions in the new design, to make it more likely to succeed.

Leave Your Desk to Learn More

Exploring the real world to learn from and empathize with customers is a valuable way to spend research time. Your research doesn't need to be overly formal, designed by an agency, or academic to be effective. It just needs to answer your research question(s) in an unbiased way. In fact, most research participants prefer to feel like they're having a chat with a friendly and curious person vs. a lab rat being studied by scientists. Good ways to get one-on-one time with your current or potential customers are usability testing, shadowing, "dogfooding," and joining the sales or service team(s). What matters most is that you genuinely get to know the people you serve and understand what they need.

EXPERT ADVICE FROM THE FIELD

MICRO SURVEY DESIGN WITH ELS AERTS

The best time to run a micro survey is when the customer enters the site or right after they complete an important task (i.e., making a purchase, signing up for an email list, etc.).

For entry surveys, some good questions to ask are, "Who are you?" and "What is your purpose for visiting this website or app today?" At first, leave these questions open and allow people to type in whatever they want.

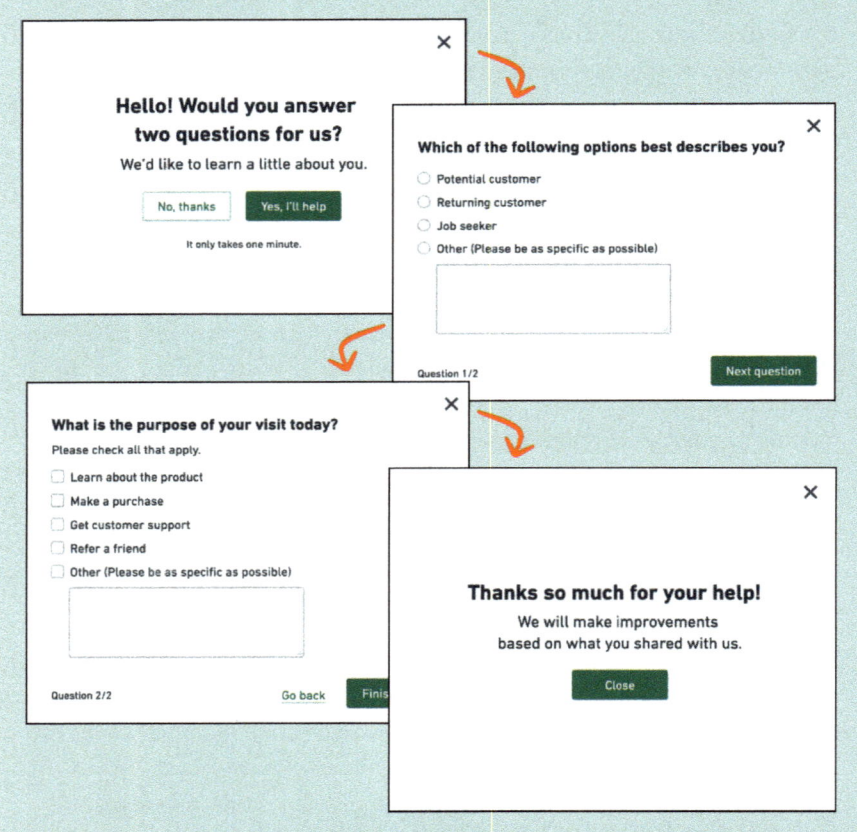

FIGURE 2.5 A basic entry survey flow to learn who visits your product or site and what they hope to do. Start with an open question and then use that information to ask a refined closed question.

This will help you make a "smart list" for a closed-question follow-up. Learning who your customers are and why they are there adds context to analytics data that you can't get anywhere else (Figure 2.5).

For a post-task completion survey, I like to ask, "What almost stopped you from completing your purchase today?" This helps because:

1. **The respondents are customers.** That means the people answering the question are in at least one target market, which means their feedback is likely to be useful.
2. **They've already taken action.** This means they've got momentum, which makes them more likely to fill out the survey than at other times. It can increase your response rate.
3. **They're likely to remember.** Human memory gets worse with time. When they fill out a survey when the experience is fresh, it's easier for them to remember what happened accurately.

For the highest response rate, embed your post-task completion surveys within the task's confirmation screen (Figure 2.6).

FIGURE 2.6 A basic post-purchase flow to find problems with the checkout process.

CONTINUES ➤

> **EXPERT ADVICE FROM THE FIELD**
>
> CONTINUED ▶
>
> These targeted questions uncover the "why" behind interesting behavior. But be sure they don't distract people from their main tasks! Don't run too many at one time, set a limit to how many people see in a session, and release them in an experiment to understand the cost of the data collection.
>
> **Els Aerts**
> **Principal researcher at AGConsult**
>
> *Els has been asking great questions for more than 20 years and teaches people to do the same at conferences worldwide.*
>
> *To dive deep into good survey design, check out the Rosenfeld title,* Surveys That Work, *by Caroline Jarrett.*

Usability Tests and Shadowing

Usability tests give participants a list of things to do with a researcher who guides or observes them. They are useful if customer context isn't important and if you just want to understand how easy or hard something is to do.

> **TIP** TEST WITH IDEAL CUSTOMERS WITHIN YOUR TARGET MARKET
>
> Always test with people in your target market who care about the product. For the best insights, give them real top tasks to do that are relevant to their personal situation. Hypothetical questions and imaginary situations provide you with poor insight. ■

Shadowing involves observing someone quietly and without interruption to learn from their natural behavior. It can be more insightful than a usability test because it's not scripted. To set up a shadowing session, find participants who need to do or learn about something related to a product. Steve Portigal's book, *Interviewing Users,* from Rosenfeld Media is an excellent resource to hone your interviewing skills. Check it out to help you run effective usability tests and shadowing sessions.

> **TIP** HAVE YOUR TEST PARTICIPANTS USE REAL MONEY
>
> It's best if people use their own money during a test to complete the task if they're ready to purchase. It's hard to find people in this situation. Providing vouchers as "real" money or a lifelike credit card number also helps. Tasks that lead to tangible gains or losses of people's own resources dramatically shift their behavior and mindset, which leads to very different outcomes than if the participants were treating it as a thought exercise. The lack of risk removes an important emotional layer of a product's experience. And emotions drive behavior. ■

Eat Your Own Dog Food (aka *Dogfooding*)

Dogfooding is when you use your own product as a real customer—with your own money and time at stake. This method creates real empathy because a fundamental mindset shift happens when done right. To do it, put yourself in the most realistic customer situation possible. This is easiest for business-to-consumer (B2C) products. It's possible to do it for business-to-business (B2B) products, but you'll likely need to be creative to make it work.

One example of how I've done this is through my work at Deliveroo, a food delivery app. I signed up as a delivery worker. To do this, I ordered the thermal food delivery bag, tried on the uniform, and hopped on my bike. I experienced the stress of having my phone battery die mid-delivery, the struggle of finding an apartment tucked away in a side street, and the frustration of a jacket hood flopping in front of my eyes in the rain.[3] I also felt the dopamine hit when I saw my earnings grow after each delivery. These problems and delights—along with their accompanying emotions—take on a different life when you experience them first-hand.

Join the Sales and Service Teams

Sales and customer service teams have regular contact with customers. They hear first-hand people's complaints, objections, and emotions. They know almost intuitively what the biggest opportunities and problems are. You should foster good relationships with these front-line workers and facilitate regular communication between product teams and sales and service teams. The closer these relationships are, the more likely that product teams will work on problems that impact business metrics.

---

3. Shout out to the kind folks who let me charge my phone in their lobby so I could deliver my last order. TALK ABOUT STRESSFUL!

After you finish your research, the data needs to be turned into useful information. One way to do this is to create an "experience landscape."

Paint a Picture with Data Points

An *experience landscape* is a user journey map enhanced with information, images, links, and videos. Whereas a user journey map captures general steps and information, an experience landscape illustrates not only high-level context, but also details of how customers behave and why. To make one, you combine both pixels and data points into one big picture. These "dots" are the medium to create a visual representation of what customers see and do within the digital space. An experience landscape is like a pointillist painting. Pointillist paintings take dots of color that individually are meaningless, but when put together become a clear picture. This is much like how an individual data point is a meaningless fact without the context around it (Figure 2.7).

SOURCE: HM SAFFER II, USED WITH ARTIST'S PERMISSION.

FIGURE 2.7 HM Saffer II's painting, "A Sunset Path," is made of many small colored dots. When you look at each dot individually, they are meaningless. When you look at all the dots together, the picture becomes clear.

STEP 1: GATHER CONTENT.

To make an experience landscape, start by taking screenshots and collecting all physical and digital artifacts that accompany the process (e.g., push notifications, triggered emails, verification letters that arrive through the mail, service calls, etc.) of the product you're visualizing.

STEP 2: GATHER DATA.

Next, gather as much relevant data related to the screenshots and artifacts as possible. Examples are:

- Conversion and usage data
- Page load time
- Form abandonment rate
- Survey data
- User research videos and screen recordings
- Number of errors
- Customer service (CS) data
- Voice of customer (VoC) data
- And so on

STEP 3: PUT IT ALL TOGETHER.

Once you have all the necessary content and data, centralize it into one comprehensive visual document to tell the customer story. Use arrows to visualize how and where different screens and pieces of content relate to one another or get triggered from. It should look like a big system diagram when you're done—not a tidy, linear user journey map. Digital whiteboarding tools are good for capturing these complex relationships and flows (see Figure 2.8).

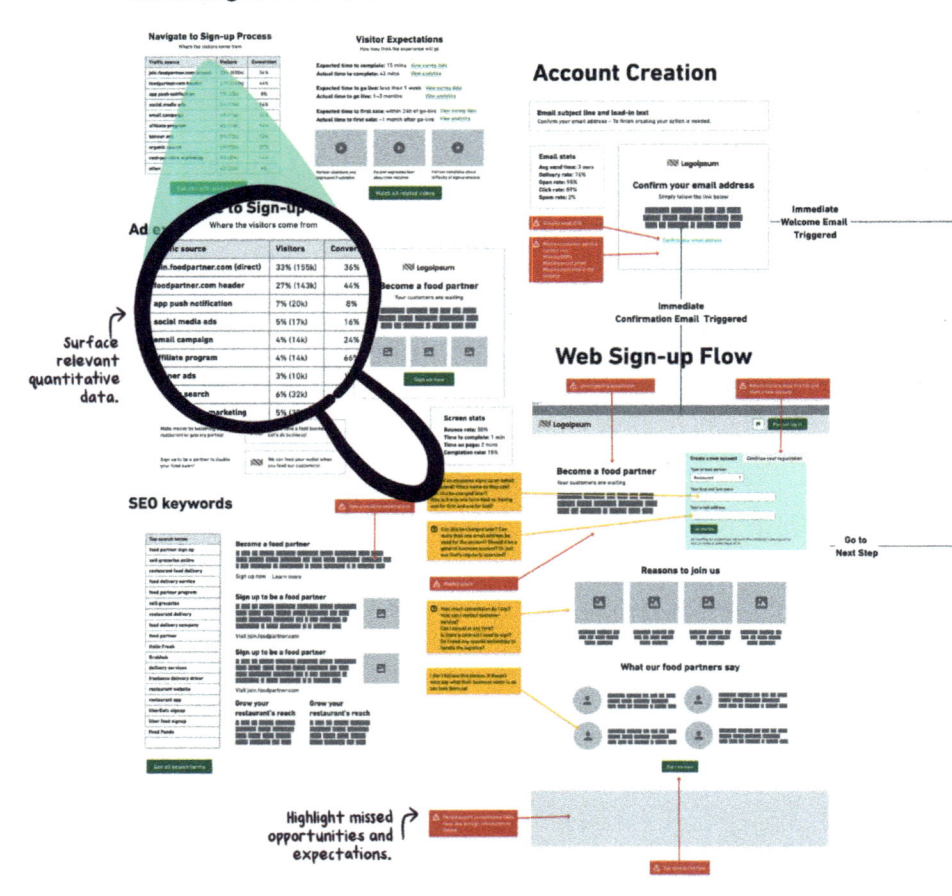

FIGURE 2.8 An experience landscape combines quantitative and qualitative data that is shown visually in context within the product. It's a high-level view of the product and how customers experience it.

TIPS

- It doesn't need to look pretty, the value comes from making it.
- Link out to related qualitative research and quantitative data.
- Use it to help others understand the holistic customer experience.

Welcome & Engagement Flow

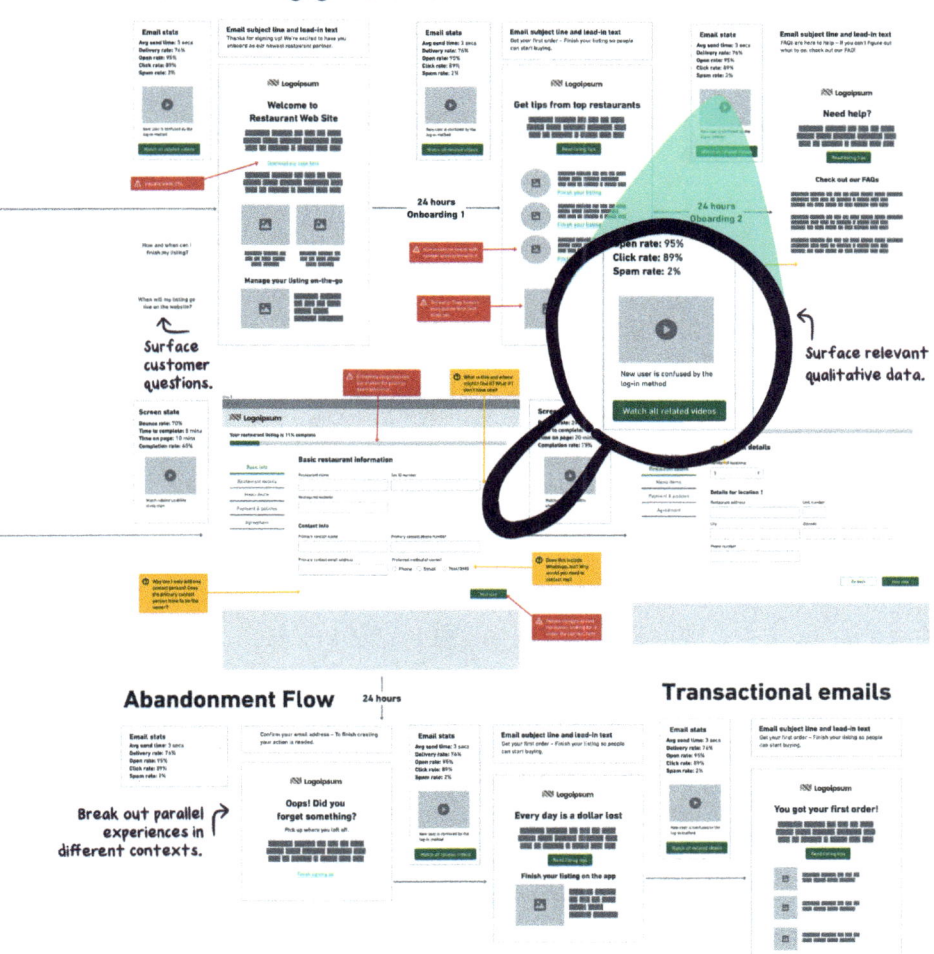

Surface customer questions.

Surface relevant qualitative data.

Abandonment Flow

Break out parallel experiences in different contexts.

Transactional emails

THE UNDERSTAND PHASE: UNCOVERING IMPACTFUL INSIGHTS · 39

Add screenshots of each step in the process or screen in the product. Connect the steps and screens with arrows and enrich the images with relevant data. For even more value, layer in your own observations, quotes from users, usability issues, and more. The more information, data, and images added, the clearer the experience landscape becomes (Figure 2.9).

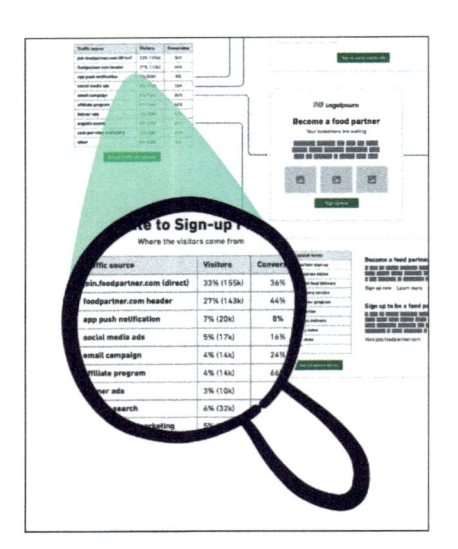

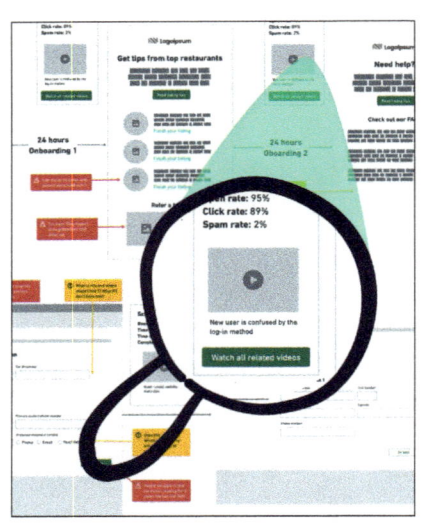

FIGURE 2.9 Visual, quantitative, and qualitative data all come together in an experience landscape. These data points "paint a pointillist picture" of the customer's experience.

STEP 4: USE IT, SHARE IT, AND KEEP IT ALIVE!

At the end, the shareable board tells the user's journey through the product. Seeing each step, task, and screen with data infused into one large picture is the closest thing to observing body language on the internet. Once the whole picture is visible, it's easier to identify the main customer problems. From there, ideas about how to address them get written down into testable hypotheses. Creating and sharing an experience landscape rallies people to fix big problems together. It breaks down silos to enable collaboration across the customer's end-to-end experience.

> **TIP** MAKE YOUR EXPERIENCE LANDSCAPE
> COLLABORATIVE AND MEMORABLE
>
> To get people to engage with its creation, ask people to share screenshots, research, questions, and data to include in the landscape. This gets them invested in its success. Then breathe life into it by telling stories about what you learned while creating it. Bonus points if you can make some stories about the experience funny. Laughter helps people remember! ∎

After you've dug into research and visualized your product's experience, your head should be *bursting* with insights. These insights are the birthplace of your hypotheses, which brings you to the next step in the Conversion Design process: the Hypothesize phase.

The Important Bits

- The first phase in the Conversion Design process, the Understand phase, is to uncover and understand customer problems through the lens of business goals.
- Start your research with a well-formulated question. Foundational ones to know are exploratory, descriptive, explanatory, and relationship-based.
- Use Christian Rohrer's "Landscape of User Research Methods" to pick the research type you need to answer your research question.
- There are two kinds of research outcomes: qualitative data and quantitative data. Both types of data are equally important, and it's best to combine both.
- Experimenting to learn helps you understand the importance of each element on a page, especially if you need to make a bigger design leap.
- Take all the data and give it context with a Customer Experience Landscape to turn it into new knowledge to build hypotheses from.

CONVERSION DESIGN PROCESS

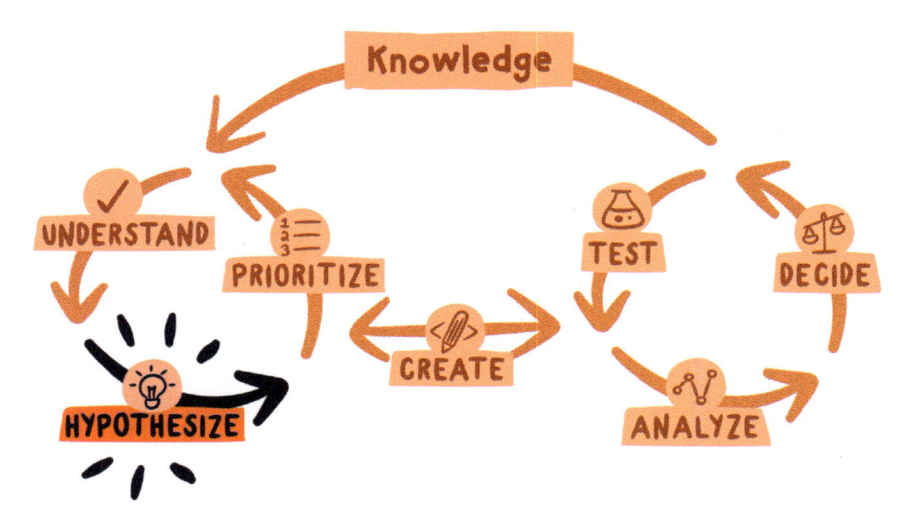

Knowledge

UNDERSTAND → HYPOTHESIZE → PRIORITIZE → CREATE → TEST → ANALYZE → DECIDE

"We are trying to prove ourselves wrong as quickly as possible, because only in that way can we find progress."

– Richard Feynman

CHAPTER 3

The Hypothesize Phase: Think Clear, Logical Thoughts

THE POINT

To measure impact accurately, use the scientific method to structure your process and thoughts. Then design your experiments properly to get reliable data for optimal decision-making.

Back in 1919, a guy named Ronald Fisher got a new job. He was hired to work at Rothamsted research station in London. His task was to look at decades' worth of agricultural experimental data and then report back on his findings. Between analyzing experiments, he'd drink tea in the break room. One day, he made a cup of tea for his coworker, Dr. Muriel Bristol.

First, he set the kettle. Then he poured a splash of milk into a mug. And finally, he filled the mug to the top with piping hot tea.[1,2]

Bristol looked at the tea with disgust and said, "I'm not drinking this."

"Why not?" asked Fisher.

"Because you cannot pour the *tea* into the *milk*. You must always pour the *milk* into the *tea*," she explained. "Pouring tea into the milk ruins the taste."

Fisher thought for a minute and then he declared, "The order makes absolutely no difference."

At this point, a coworker named William Roach overheard the argument and stepped in. "Let's run an experiment to see who's right," he proposed.

Bristol and Fisher, both scientists, agreed to the challenge. Fisher made eight cups of tea and presented them to Bristol in a random order—four with the *tea* poured into the *milk* and four with the *milk* poured into the *tea*. Surprisingly to Fisher, Bristol identified all eight cups of tea correctly. He quickly calculated the probability of her guessing every single cup of tea right. It was only a one-in-seventy chance. This made Fisher believe that Bristol could indeed detect some kind of difference.[3]

To stress test his confidence, he calculated the probabilities of her guessing only six out of the eight cups correctly. In that scenario, the odds dropped from one-in-seventy to one-in-four. He realized that this massive change in confidence between guessing *all* correctly vs. guessing *most* correctly meant the small sample size of eight put a disproportionate amount of weight on individual cups of tea.

---

1. Sam Kean, "Ronald Fisher, a Bad Cup of Tea, and the Birth of Modern Statistics," *Distillations Magazine*, Science History Institute Museum & Library, August 6, 2019, www.sciencehistory.org/stories/magazine/ronald-fisher-a-bad-cup-of-tea-and-the-birth-of-modern-statistics/
2. Peter Lynch, "How a Tea-Tasting Test Led to a Breakthrough in Statistics," *The Irish Times*, September 5, 2019, www.irishtimes.com/news/science/how-a-tea-tasting-test-led-to-a-breakthrough-in-statistics-1.3998786
3. Bristol was right. It was eventually discovered that a chemical change takes place when hot tea hits the mass of milk. The temperature shock forms globules in the milk fat, which subtly changes the taste.

Fisher and Bristol's "tea test" sparked a new way of thinking about experimental design by introducing these new ideas:

- **Randomized samples** with a control to compare the change against
- **A skeptical mindset** to sway you from your default belief (called *null hypothesis testing*)
- **Statistical analysis** to understand the probability of the result (called *ANOVA*—short for *analysis of variance*)

Fisher applied these ideas to his day job. After poring through thousands of documents, he came to a shocking conclusion: all the data was garbage. Nothing could be learned from it due to poor experimental design. He soon realized that this was a common problem in the field of research, so he wrote two books to propose the solution. These books provided the world with the foundations of modern experimental design: *Statistical Methods for Research Workers* and *The Design of Experiments*.

Although Fisher's methods drastically improved his field of research, experiment design quality in online testing is where most people go wrong. So, dust off your science pants! It's time to dig into the scientific method and Fisher's experimental design approach so that you can collect data that you won't have to toss in the trash.

The Modern Scientific Method

The modern scientific method Fisher designed is a structured way of asking questions to get reliable answers. It has six steps:

1. **Make an observation** while doing your research.
2. **Form your null and alternate hypotheses**, which are theories inspired by your research. These are the starting points for further investigation.
3. **Make a prediction**, which is what you think will happen if the alternate hypothesis is true.
4. **Run a test** to collect reliable data.
5. **Analyze the results with statistics** to learn if the data disagrees with your initial belief (null hypothesis) or not.
6. **Form an opinion by considering all the evidence** collected up to that point to come to a balanced conclusion.

Conversion Design is based on Fisher's modern approach and the practice of design thinking. Design thinking is a creative approach to solving problems for people. It focuses on understanding people and their needs to generate new ideas. To account for design thinking in Conversion Design, one step is added to the scientific method: "Define the problem." It slides in between, "Make an observation," and "Form a hypothesis," as shown in Figure 3.1.

Fisher's Scientific Method

1. **Make an Observation** Through Research
2. **Form a Hypothesis** a.k.a. a "Testable Idea"
3. **Make a Prediction** From Your Testable Idea
4. **Run a Test** Based on Your Prediction
5. **Analyze the Results** To Learn From Them
6. **Form an Opinion** Based on the Results

Conversion Design's Scientific Method

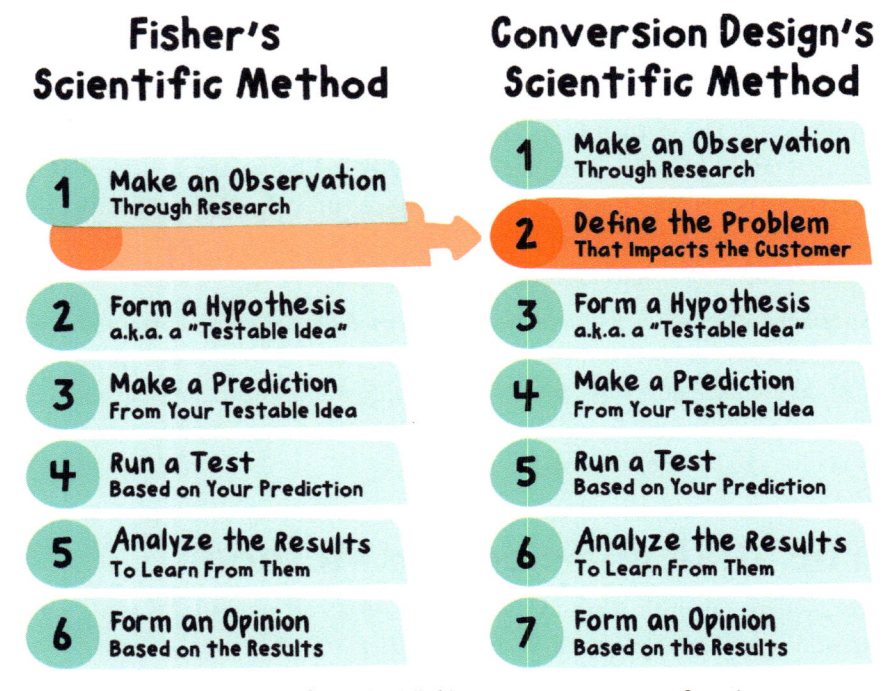

1. **Make an Observation** Through Research
2. **Define the Problem** That Impacts the Customer
3. **Form a Hypothesis** a.k.a. a "Testable Idea"
4. **Make a Prediction** From Your Testable Idea
5. **Run a Test** Based on Your Prediction
6. **Analyze the Results** To Learn From Them
7. **Form an Opinion** Based on the Results

FIGURE 3.1 Fisher's scientific method (left) gets an extra step to reflect design thinking in Conversion Design's scientific method (right).

This additional step shapes your hypothesis and your prediction in order to be customer-centric. Your prediction, typically written in an if/then format, should be a solution to the customer problem. But before you can propose a solution, you need to define your hypotheses.

The Null and Alternate Hypotheses

Null-hypothesis testing, Sir Ronald Fisher's innovation from the tea experiment, divides a hypothesis into two camps.

1. **The null hypothesis.** This is the default belief or currently accepted theory. Fisher's null hypothesis in the "tea test" was that the order the milk mixed with the tea *did not* matter. The null hypothesis is also sometimes shortened to the acronym *H0*.

2. **The alternate hypothesis.** This is the new idea that challenges the null hypothesis. It was Bristol's claim that the order in which milk got added to the tea *did* matter. It's sometimes called *Ha* or *H1*.[4]

Table 3.1 breaks down the two types of hypotheses for you.

TABLE 3.1 THE TWO TYPES OF HYPOTHESES

NULL HYPOTHESIS (H0)	ALTERNATE HYPOTHESIS (H1, HA)
The belief behind the default situation. The world without any change in it. It is the "A" in an A/B test.	The belief behind the change introduced in the experiment. The world with a change in it. It is the "B" in an A/B test.

Contrary to popular belief, the idea behind hypothesis testing is *not* to prove the alternate hypothesis *true*.[5] It is to be extremely skeptical of the alternate hypothesis and assume that the null hypothesis is true. You only *reject* the null hypothesis when very convincing evidence shows that you should. In this way, it's like stubbornly sticking to what you know.

---

4. When I first learned about H0, Ha, and H1, I wished these number/letter combos were just some fun internet way to laugh and say "hi." Unfortunately, I was wrong.
5. I thought this was the case for a super-duper long time. So, if you thought this, too—it's fine.

LEARNING TO CHANGE MY MIND

Back in the day, I used a regular old toothbrush. It was fine. However, my friend, Lucas, told me that electric toothbrushes were better.[6] I thought about the suggestion with three ideas in mind 1) They are more expensive, 2) Lucas loves gadgets, and 3) I'm stubborn.[7] I assumed an electric toothbrush was a pricey doohickey that would clutter up my sink. Naturally, I ignored his advice and stuck with what I had.

Then one day, my dentist told me, "You should get an electric toothbrush." And it made me think: At that point, two people—both smart and without an ulterior motive—had told me that electric toothbrushes were better. So, I reconsidered my default position. That night, I bought an electric toothbrush to test against my regular one. I was going to learn if I should change my mind.

In this case, my null hypothesis was "regular toothbrushes work as well as an electric toothbrush." The alternate was "electric toothbrushes work better than regular ones" (Table 3.2).

---

6. This is "toothbrush Lucas." His real name is Lucas Silenski-Cahill. I have many Lucas/Lukas friends in this book. Brace yourselves.
7. Why, yes. I *am* a Taurus.

TABLE 3.2 TOOTHBRUSH EXPERIMENT HYPOTHESES

NULL HYPOTHESIS	ALTERNATE HYPOTHESIS
Regular toothbrushes *work as well as* electric toothbrushes.	Electric toothbrushes *work better than* regular toothbrushes.

Unless I could be convinced that electric ones were *really better*, I was—*for sure*—returning it.

It's this idea of "sticking with the default unless proven otherwise" that's the spirit behind null hypothesis testing. Richard Feynman, a Nobel award-winning physicist and artist, summed up null hypotheses best when he said, "We are trying to prove ourselves wrong as quickly as possible, because only in that way can we find progress."

Anyway, after a few nights of brushing with an electric toothbrush, I was convinced that the electric toothbrush was indeed better.[8] I rejected my null hypothesis… and I saved a trip back to the store, too!

---

8. My success criteria were how slippery it felt when I ran my tongue across my teeth. Generally, speaking—this isn't a good form of measurement for most experiments. Ha!

Hypothesizing in a Business Context

How does all of this talk about tea and toothbrushes influence how you should do your job? Well, imagine you work for an e-commerce company, and you're watching someone shop for a shirt online. The available sizes are small, medium, and large. The shopper doesn't know if they fit in a medium or a large. So, they order both. "I'll just send back the one that doesn't fit," they say as they check out. Suddenly, you realize: You just observed a customer problem!

> **CUSTOMER PROBLEM**
> People don't know what size to order, so they order the same product in multiple sizes.

You hypothesize that customers need sizing information to make a confident purchase. From this, you form your alternate hypothesis, a testable idea. It is a potential solution to the customer problem you observed.

> **ALTERNATE HYPOTHESIS**
> Sizing information *has an impact* on customer confidence and order decisions.

Your prediction, which is based on your alternate hypothesis, gets written in if/then format.

> **HYPOTHESIS PREDICTION**
> *If* sizing information is available, *then* people will be more confident in their decision and not over-order.

Next, you state the null hypothesis. This captures the idea that the current experience has no sizing information available. This is your default belief that will be challenged.

> **NULL HYPOTHESIS**
> Sizing information *has no impact* on customer confidence and order decisions.

(continues on page 52)

EXPERT ADVICE FROM THE FIELD

CONNECTING CUSTOMER PROBLEM STATEMENTS TO TEST IDEAS WITH SHIVA MANJUNATH

Hypotheses inspired by research and driven by customer problem statements are more likely to succeed than those that aren't. That's why I created "The Problem Statement Driven Hypothesis Blueprint" shown in Figure 3.2.

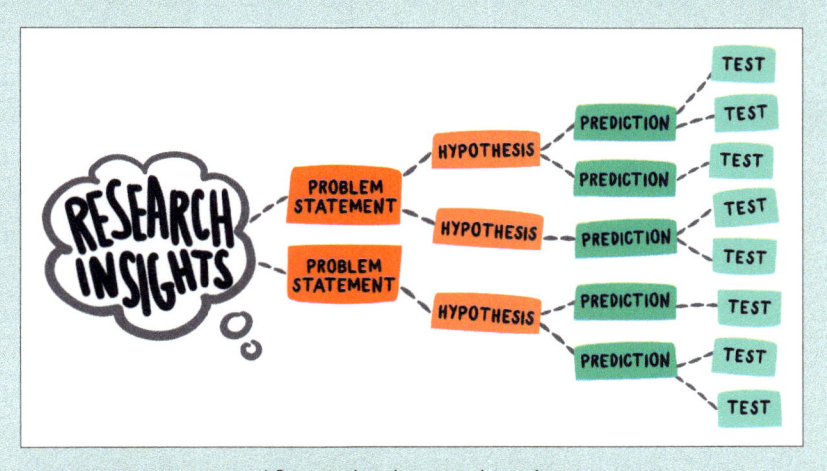

FIGURE 3.2 An optimal flow to develop your hypotheses.

This blueprint is designed to help teams more easily make the connection between research insights, the customer problem, the hypotheses and predictions for how to solve the customer problem, and all the specific test variants that could be run.

This blueprint grounds all your A/B test in customer research, and it uses "customer problem statements" as the bridge. You can (and should) link these customer problem statements back to business goals. This way, when you present your experiment plan to leadership, it's easier to agree on the customer problems that should be solved instead of getting lost in a sea of unanchored A/B test ideas.

CONTINUES ➤

> **EXPERT ADVICE FROM THE FIELD**
>
> CONTINUED ➤
>
> **Shiva Manjunath**
> **Host and founder of the Experimentation Podcast *From A to B*.**
>
> *Shiva's been running experiments for more than 10 years. Find him on LinkedIn where he shares hilarious experimentation memes and useful A/B testing advice to make your job less hard.*

So far, your hypotheses only mention customer confidence and ordering behavior. But customer confidence isn't measurable in an unbiased way. And there's no clear thread back to a business goal. This is where you have to be creative. You need to figure out what to measure that could be a good "proxy" for customer confidence. A *proxy* is someone or something that has the authority to represent someone or something else. For example, if you're not able to appear in person to sign a contract, you could give someone the right to sign on your behalf. This person represents you and your interests as your proxy.

You remember that the customer said they would return the size that didn't fit. This is a simple number: The company's return rate. Returns are bad for both the customer and the business because they waste the customer's time, and they cost the business money.[9] Therefore, "return rate" would make a good proxy for customer confidence in your experiment. So, there you have it! From one casual customer comment, you developed all the foundational logic you need to design an experiment that's good for both customers and the business.

Types of Experiment Variables

As you learned in Chapter 2 (in the section "Ask Questions to Collect Data"), a hypothesis is a relationship-based research question. It's an educated guess that attempts to explain the relationship between two or more "things."

---

9. They're also terrible for the environment. Remember: The planet and natural world are also key stakeholders in your experiments.

These *things* are called *variables*. The alternate hypothesis in our example, "Sizing information impacts customer confidence," has two variables to examine: sizing information and the return rate (the proxy for customer confidence). This example (shown in Table 3.3) explains the relationship between showing sizing information (the independent variable) and the return rate (the dependent variable). The bold words are the variables.

TABLE 3.3 THE EXPERIMENT VARIABLES IN THIS EXAMPLE

Sizing information (Independent Variable)	impacts	**return rate.** (Dependent Variable)

The first thing, "sizing information," is the independent variable. This is what you will *intentionally change*—or manipulate—in your experiment to see what happens. The second thing, the "return rate," is the dependent variable. This is what you *predict will be affected* because of the change. It's what you measure to learn if there is a relationship between the two variables. Figure 3.3 illustrates a cause/effect relationship ("causality") between the independent and dependent variables that you're trying to keep "clean" ("isolated") and measure in your experiment.

Cause
The Thing That's Manipulated
(Independent Variable)

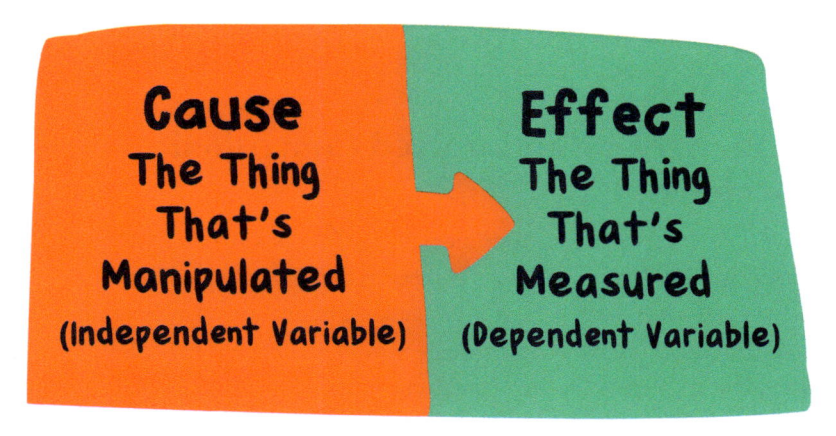

Effect
The Thing That's Measured
(Dependent Variable)

FIGURE 3.3 Variables are things you watch to understand how one thing impacts another.

The problem is that bias and other unrelated factors—known as *confounds*—can confuse the situation and throw off the experiment results. In other words, how can you be sure the change you made actually caused the result you got? Potential confounds in this situation could be things such as sales at competing stores, minimum purchase return policies, customer location, and customer type (e.g., new vs. returning). Confounds make determining causality difficult, if not impossible. To understand the concept of confounds, imagine that you couldn't randomize your sample in the shirt-size example. Imagine your variant (the one that shows sizing information) had more returning customers than new customers. This could be a problem because returning customers might be familiar with the product sizes from past purchases. So, sizing information could be less important to these customers. In this case, customer type is a confounding variable because it might impact the result of the experiment. The effect of customer type must be removed so you can be certain about what causes any change you might observe. But how can you remove confounds from your experiment? By *controlling* them.

Controlling confounds is when you carefully spread the effect of a confound equally between samples. Once the confound's effect is neutralized by spreading it equally across the variants, it's called a *controlled variable*. Controlled variables are confounds that will not affect the result because the confound's impact is removed through randomization. Randomization is a highly reliable way to spread the confounds equally between the treatment and control to remove their effect. That's right—the potential impact of confounds is completely removed when you randomize your sample properly. Randomization is the secret sauce of A/B testing.

The Birth of a Randomized Controlled Experiment

Randomization is when each experiment participant has an *equal chance* of being assigned to either base or variant. This means that when a sample is randomized, neither base nor variant is advantaged or disadvantaged by anything other than *the only true difference between them*—your independent variable. Once the independent variable is the only consistent difference between the two variants because of randomization, it's called a *randomized controlled experiment*.

A 50/50-split A/B test is the simplest type of randomized controlled experiment. It acknowledges all the complex relationships between many things that could impact a result with systems thinking. Then, through randomization, it creates a simple linear relationship that you can learn from. Table 3.4 summarizes the four kinds of variables to account for in your experiments.

TABLE 3.4 FOUR KINDS OF EXPERIMENT VARIABLES

INDEPENDENT VARIABLE (CAUSE)	DEPENDENT VARIABLE(S) (EFFECT)	CONFOUNDING VARIABLE(S) (MUST BE CONTROLLED)	CONTROLLED VARIABLE(S) (MUST STAY THE SAME)
The thing that gets changed to see if there's an impact on the dependent variable(s).	The thing(s) that gets measured to understand the impact of the independent variable.	Things that can impact the independent and dependent variable and may influence the results of the test. The effect of confounds must be controlled through randomization.	Everything else that needs to stay the same or be "controlled" to be sure any difference measured in the dependent variable is caused only by the independent variable.
Example: Sizing information vs. no sizing information.	Example: Return rate (proxy for customer confidence).	Examples: Coupon codes, return policies, location, type of customer (new vs. returning), etc.	Examples: Any of the confounds that have had their effect neutralized through randomization.

AND POOF—THEY'RE GONE!

Every once in a while, statistics give me an existential crisis. One day, my brain couldn't fully grasp how randomization ensures that the *only thing* impacting the result of my test was the change I made. In a panic, I called my friend, Lucas, to frantically "word vomit" half-formulated questions at.[10]

"Statistics Panic Attack Hotline," Lucas said as he picked up the phone. "What's breaking your brain today?"

CONTINUES ▶

---

10. This is not "toothbrush Lucas." This is a new Lucas named "Lucas Bernardi." I also affectionately call him "cake Lucas." But you have to read Chapter 6 to learn why.

CONTINUED ➤

"Randomization." I replied. "*How* can I trust the cause/effect relationship?"

"Easy," he declared.[11] "Up until Fisher introduced randomization, scientists were typically handed a pile of data where clear relationships were hard to find. They had to find all the possible things that could influence the results and do complex analyses to manually remove as many of them as possible. But when you design your experiment upfront to include a randomized sample, the problem of confounds literally just disappears. Poof—gone! When you randomize your sample properly, you *don't have to worry about confounds at all anymore*. And when you combine randomization with large sample sizes, then you can worry about confounds *even less*."

"How can I worry about something even less than not at all?" I asked.

"That's the point!" he laughed. "Statisticians figured out how to randomize for you, so you don't have to worry about it. What you *do* want to worry about, though, is making sure the randomization worked OK," he added. "So, randomization can go wrong?" I asked. "How can I make sure my randomization is OK then?"

"Easy," he declared.[12] "You can learn if your randomization is working OK by checking for a sample ratio mismatch (SRM)."

"Oh great," I said. "Science words with an acronym—this is going to be fun."

Then Lucas explained to me all about randomization and SRMs, which is why the next section is called, "Randomization and Sample Ratio Mismatches (SRMs)." In it, Lucas and I made a collaborative attempt to explain randomization and SRMs in a (somewhat) approachable way.

---

11. For the record, I did not think it was "easy."
12. Again—I did not think it was "easy."

Randomization and Sample Ratio Mismatches (SRMs)

When most people think of "random," they think about board games and dice. A key feature of rolling a die is that it doesn't matter who rolls it. The chance of rolling 1, 2, 3, 4, 5, or 6 is the same for all participants. Whether a player rolls a 2 or a 5 has nothing to do with anything—not the person's sitting position, how much they believe they can control the die, nor what they had for breakfast that day. If it were possible that one person could somehow control the die, or if the die somehow helped some participants while hurting others, the whole game would be ruined. Randomized controlled experiments rely on a similar key feature: the chance of any person being placed in the treatment or control is exactly the same for everyone. Nothing at all about the person or the situation influences what variant participants are assigned to.

You start by setting the ratio you want your sample to be split into in your A/B testing tool (or "experimentation platform").[13] Your testing tool does the randomization and reports the result for you. The ratio you choose can be whatever—50/50, 99/1, 90/10—it doesn't actually matter. The probabilities do not need to be the same between treatment and control, but they do need to be the same for everyone. When the split happens according to what you told your testing tool to do, randomization successfully gets rid of all the confounds and isolates your independent variable.

In board games, ensuring randomness with a fair die is quite straightforward. You just need to make sure that the die rolls a few times (since dice are very hard to control once out of our hands). However, in randomized controlled experiments, proper randomness can be a lot harder to achieve. The stakes are high in online experimentation because if your tool doesn't get randomization right, the results are completely unreliable. So, it's a good idea to make sure that your randomization is OK with "sample ratio mismatch (SRM) check."[14]

A sample ratio mismatch (SRM) happens when the number of participants placed in the treatment compared to the control is extremely unlikely to happen

---

13. If you don't have an A/B testing tool yet, check out the A/B Testing Tool Comparison website by Speero. Find it at www.erindoesthings.com/resources or at https://speero.com/ab-testing-tools
14. Good A/B testing tools check for SRMs and notify you of one automatically. Get a good tool.

if all users truly had the same assignment probabilities. An SRM is a symptom of an underlying problem with your data quality. Basically, an SRM tells you that something went wrong with how experiment participants were randomized.

For experiments where users are placed in the treatment or control with a 50/50 chance, an SRM happens when the number of people in one variant is suspiciously bigger than the number of users in the other variant. "Suspicious" means that the difference is big enough to make you doubt that the chance people had to be the control group, or a treatment was actually 50/50. Because of randomness, you will almost never observe the same number of users in each variant. But this is expected, and it's nothing to worry about. You should only get worried if the difference is big enough that it's statistically unlikely to happen.[15]

When you observe an SRM, it's likely that some people had a higher chance of getting assigned treatment or control. This breaks the randomness, and ultimately, invalidates your experiment result.

How to Spot an SRM

Detecting an SRM isn't terribly hard. You simply look at the number of participants assigned to the control vs. the treatment. Then you calculate the probability of that split happening if the split were truly random. If it were truly random, the probability of the observed split happening should be statistically likely. If the randomization isn't OK, then the split you observe would be statistically unlikely. Luckily, you don't have to calculate the probability yourself. You can use the Sample Ratio Mismatch Checker on Lukas Vermeer's website shown in Figure 3.4.

To understand how this all works, imagine you run a 50/50 experiment with 1,000 participants. You should expect to observe about 500 people in each variant. However, imagine that you observe 600 in the treatment and 400 in the control. The probability of observing a difference of 20 users between variants is about 50%. But the probability of observing a difference of at least 200 users is approximately 1 in 3.6 billion. So, observing a split of 600/400 is very suspicious. This unlikelihood tells you that something went wrong in the assignment process, and it suggests that your randomization is broken.

---

15. You will learn about what's "statistically likely" and "statistically unlikely" in Chapter 7. So, don't worry if you don't quite understand what this means yet. I suggest that you come back and read this section again after you read Chapter 7 to help solidify your understanding.

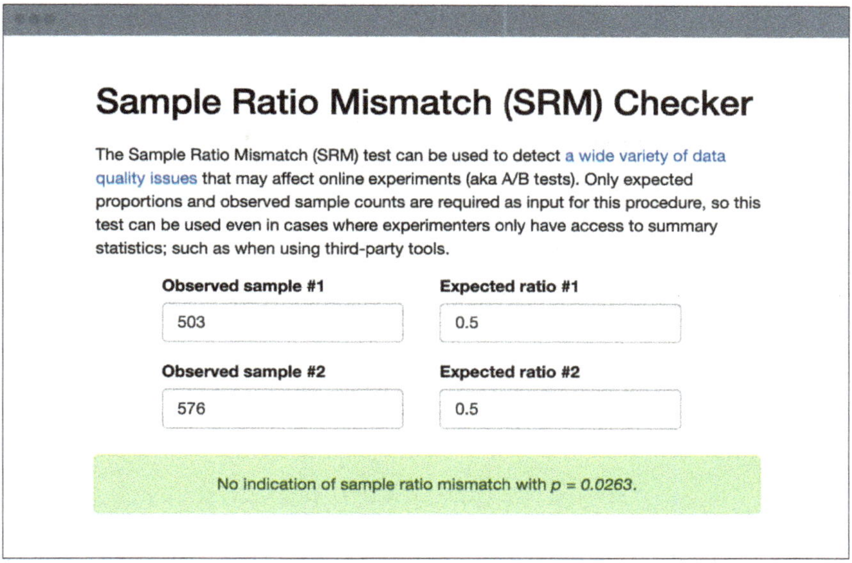

Sample Ratio Mismatch (SRM) Checker

The Sample Ratio Mismatch (SRM) test can be used to detect a wide variety of data quality issues that may affect online experiments (aka A/B tests). Only expected proportions and observed sample counts are required as input for this procedure, so this test can be used even in cases where experimenters only have access to summary statistics; such as when using third-party tools.

Observed sample #1	Expected ratio #1
503	0.5

Observed sample #2	Expected ratio #2
576	0.5

No indication of sample ratio mismatch with $p = 0.0263$.

FIGURE 3.4 Lukas Vermeer's Sample Ratio Mismatch Checker at www.LukasVermeer.nl/srm/microsite/.

What to Do If You Spot an SRM

So, what should you do if you observe an SRM? Check your experiment setup for bugs, fix the bug(s) if you find any, and run the experiment again. If you don't find any bugs, rerun the experiment anyway. If you observe an SRM again, you missed a bug in your experiment setup. If you observe no SRM during the fresh run of the experiment, the data is likely OK. A common situation where SRMs happen is with fancy JavaScript event tracking. This type of tracking typically gets triggered "on-view" (when a certain thing becomes visible on the screen) or "on-click" (when a user interacts with a link, button, image, icon or whatever). Be *very* careful when you use front-end JavaScript tracking (or "client-side" tracking) vs. back-end tracking (or "server-side" or "software development kit (SDK)" tracking). Back-end tracking tends to be more reliable because it gets triggered immediately upon the page load request vs. whenever the JavaScript happens to load.[16]

---

16. If you want to read more about types of experiment tracking, check out www.convert.com/blog/a-b-testing/client-side-testing-vs-server-side-testing/

The point of this section is that data with an SRM is not trustworthy, and you shouldn't use it to make decisions. Understanding SRMs, though a foundational concept to collect reliable data, is not basic A/B testing

Why Randomization in Experimentation Is Important

Confounds could mess up your result.

Confounds are things that are likely to influence the results of the test but are not the focus of the study. They are factors, which are not your independent variable, that can confuse the situation and throw off the results.

Examples are: user type (new or returning), customer location, traffic source, etc.

How to Check if Your Sample Is Randomized OK

Step 1. You start with a sample.

There are 20 people in this example. Some have a particular confounding factor and some don't.

Step 2. Choose the ratio you want.

Set your testing tool to a ratio that splits the sample into the percentage you want exposed to the control vs. the percentage in the treatment group. This example shows a 50/50 split.

Step 3. Your testing tool randomly splits the sample into the ratio you want.

Your tool does the coin-flip math with an algorithm.

Step 4. After the sample is split, do a "Sample Ratio Mismatch (SRM)" check.

An "SRM check" is how you tell if your randomization worked OK or not. If you have a good A/B testing tool, it checks for SRMs, and it notifies you if one is found.

Find a list of tools with an SRM-check feature by filtering for SRM at:
https://speero.com/ab-testing-tools
Or use the manual SRM checker at: LukasVermeer.nl/srm/microsite

FIGURE 3.5 The process flow for how to check for an SRM in your experiment data.

information. To help solidify your understanding, check out Figure 3.5. It illustrates how to check for an SRM in your experiment and how the confounds behave when a sample is either properly or improperly randomized.

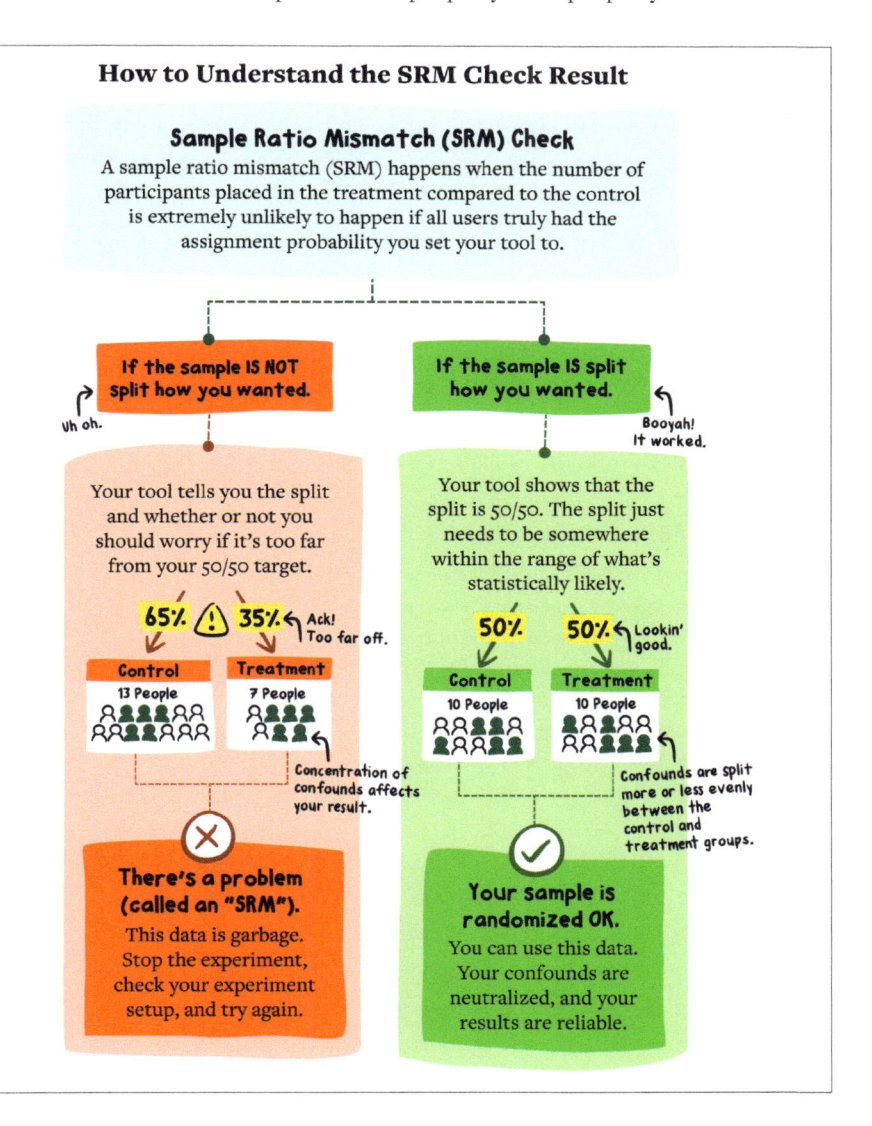

How to Understand the SRM Check Result

Sample Ratio Mismatch (SRM) Check

A sample ratio mismatch (SRM) happens when the number of participants placed in the treatment compared to the control is extremely unlikely to happen if all users truly had the assignment probability you set your tool to.

If the sample IS NOT split how you wanted.
Uh oh.

Your tool tells you the split and whether or not you should worry if it's too far from your 50/50 target.

65% ⚠ **35%** Ack! Too far off.

Control — 13 People
Treatment — 7 People

Concentration of confounds affects your result.

❌

There's a problem (called an "SRM").

This data is garbage. Stop the experiment, check your experiment setup, and try again.

If the sample IS split how you wanted.
Booyah! It worked.

Your tool shows that the split is 50/50. The split just needs to be somewhere within the range of what's statistically likely.

50% **50%** Lookin' good.

Control — 10 People
Treatment — 10 People

Confounds are split more or less evenly between the control and treatment groups.

✓

Your sample is randomized OK.

You can use this data. Your confounds are neutralized, and your results are reliable.

THE HYPOTHESIZE PHASE: THINK CLEAR, LOGICAL THOUGHTS · 61

Measuring Success

How do you go about measuring success in an experiment? Three types of metrics help tell the story: goals, drivers, and guardrails.[17] *Goals* are what you're trying to achieve. Ideally your goal adds value to both customers and the business. Most businesses have specific goals or success metrics defined by leadership. These can be used as your ultimate measure for whether a proposed change is helpful or not. If business goals aren't already spelled out, you'll have to define your own at the beginning of an experiment.

Goals should be measurable in a "reasonable" amount of time. What's "reasonable" depends on the nature of the business and the goal. If you're not able to observe the business impact you aim for on your goal within a reasonable amount of time, you need to get creative with selecting a good proxy for your real goal. Since proxies are an assumption, you must regularly assess if your proxy is truly linked to your actual goal or not. Picking the right goals is *essential* to designing the impact you aim for.

> **TIP NUMBER OF CLICKS IS A LOUSY GOAL**
>
> Product teams sometimes work to increase clicks on an important button or link. It's impossible to know if more or fewer clicks are better for the customer. Clicks in-and-of-themselves also add no real business value. When a metric is just something to measure that has questionable impact, it's called a *vanity metric*. It's just there for people to look at and feel like they're making a difference. Better goals are number of sales, calls or chats to the customer service team, or return/cancellation rate. Those goals all have some kind of value for both the customer and the business. ∎

Drivers cause or influence goals. They are easily measured within a short period of time and are known as *leading indicators* because they are predictive of a goal being achieved. For example, a driver to a complete purchase (goal metric) might be a user adding an item to their shopping cart. The customer starting the checkout process is also something that needs to happen. A sale cannot be made without an item first being placed into the cart and the checkout process being started. In this way, drivers indicate intent and lead the way to goal completion.

---

17. Ron Kohavi, Diane Tang, and Ya Xu, *Trustworthy Online Controlled Experiments: A Practical Guide to A/B Testing* (Cambridge, UK: Cambridge University Press, 2020), 90–92.

Guardrails are what you look at to make sure that no damage is done to other important things while achieving the drivers and main goal. These are sometimes known as *lagging indicators* because they are typically reflective in nature and usually take more time to measure. Some examples of lagging indicators are order cancellations or returns, customer service contacts, or repeat purchases (loyalty). Set your guardrail metrics up before you run your test to ensure that no harm is done to one stakeholder while achieving a benefit to another. Table 3.5 summarizes the types of metrics you should include in your experiment's design.

TABLE 3.5 THE THREE TYPES OF METRICS

GOAL	DRIVER(S)	GUARDRAIL(S)
The thing both the business and the customer want to achieve.	Things that need to happen before a goal can be complete.	Health metrics that shouldn't be impacted to ensure that the overall impact is net positive.
Also known as: • Classic indicator • Macro-conversion • Primary metric	**Also known as:** • Leading indicator • Micro-conversion • Secondary metric	**Also known as:** • Lagging indicator • Health metric • Monitoring metric
Examples: Number of sales, number of customer service contacts, number of demo requests	**Examples:** Items put in cart, checkout process started, sign-ups, etc.	**Examples:** Item returns, customer service contacts, canceled orders, loyalty

When you acknowledge the complex impact a change can have on your customers and business, you can make "optimal" decisions. Optimal decisions balance conflicting goals to maximize potential benefit while minimizing harm. Now that you know the basics of the scientific method and how experimentation works, it's time to pick which idea you want to test first. Next up: The Prioritize phase!

The Important Bits

- A customer-centric scientific method structures the thinking behind Conversion Design's experimentation process. Its steps are the following: make an observation, define the customer problem, form your hypotheses, make a prediction, run a test, analyze the results, and form an opinion.

- Modern experimental design has two types of hypotheses: the null hypothesis, which adds a skeptical mindset to testing, and the alternate hypothesis, which is a testable idea that challenges the null hypothesis.

- Experiments have four types of variables: independent (cause), dependent (effect), confounding (must be controlled), and control (must stay the same).

- Randomized controlled experiments, such as an A/B test, are the most reliable forms of scientific evidence. They use both systems and linear thinking to learn about potential cause-and-effect relationships.

- Proper randomization, when designed into an experiment up front and combined with a large sample size, completely removes confounds. Data scientists work hard to get randomization right, but one way to tell if something might be wrong with your randomization is to check for a sample ratio mismatch (SRM).

- To acknowledge the systemic nature of experimental results, and thus make "optimal" decisions, you should balance goals, drivers, and guardrail metrics. This helps you measure the impact of your decisions on different stakeholders over both the short- and the long-term.

CONVERSION DESIGN PROCESS

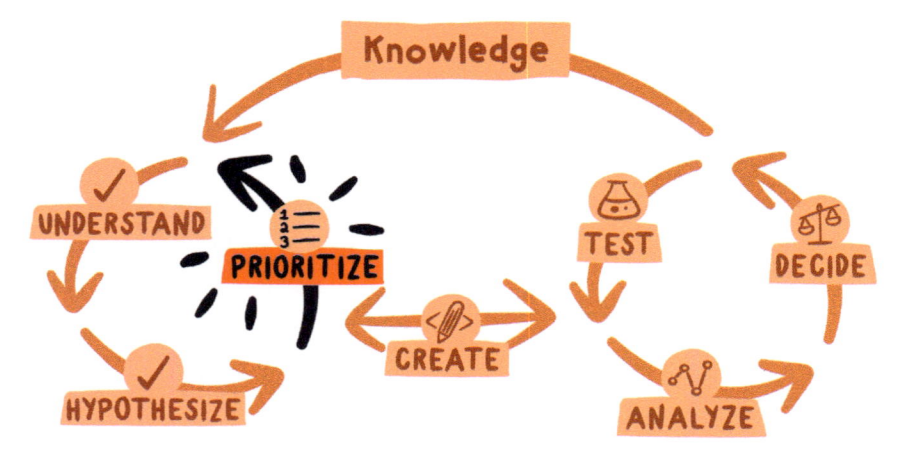

Knowledge

UNDERSTAND → HYPOTHESIZE → PRIORITIZE → CREATE → TEST → ANALYZE → DECIDE

"We think in order to act, but we also act in order to think. We try things, and those experiments that work converge gradually into viable patterns that become strategies."

— Henry Mintzberg

CHAPTER 4

The Prioritize Phase: The Work and the Workflow

THE POINT

What you work on and how you work influences how quickly you can learn from the Conversion Design process.

After the Understand and Hypothesize phases, you should have lots of experiment ideas. But—you can't do them all! That's where prioritization comes in. Ideally, you only work on ideas that deliver direct value to customers and the business. Unfortunately, you can't know what those things are until you test them. That makes prioritization an advanced form of guessing.

A general order of importance when "guessing" what to work on is:
- Build the right thing.
- Make it work.
- Make it excellent.

This order loosely reflects different stages of product maturity.

The Right Thing Made Right

Focus on building the right thing first, because the wrong thing made right adds no value, but the right thing made wrong can.

For example—imagine your product team decides to mass manufacture a pickle grabber. You don't know if there's really a market for the product or not, but you have a strong feeling that it's something most people want and need in their lives. So, you invest a bunch of money making the most beautiful and ergonomically well-designed pickle grabber ever. You launch the product with the coolest name and brand, but it flops. You did everything "right" by ensuring that it was usable and aesthetically pleasing and well-branded, but no one wants it, so it's pointless. In other words, if you make something useless that no one wants, but you make it perfectly usable, beautiful, and accessible, people still aren't going to find value in it, because nobody *needs* it.

Now imagine your product team identifies a real problem they've observed many people struggling to solve—for example, finding the right lid to the right reusable container. So, you get busy solving this problem, but you use cheap plastic, the containers are weird shapes, and it has unappealing colors. Also, the branding looks terrible. But the problem is annoying enough for people who appreciate your plastic solution, that they buy the product anyway—ignoring the weird colors, atypical shapes, and crappy logo. You made the right thing in a way that's not done *right* from a presentation and usability perspective. In other words, if you make a product that people actually need and want, the problem it solves is big enough that people will overlook usability and accessibility issues and deal with the struggle because it makes their lives better.

The previous scenarios emphasize the importance of product-market fit in experimentation. Product-market fit happens when a company creates something valuable enough that people "invest" in it. Customers have various forms of currency to invest. Money, time, and attention are a few. Each of these is a form of business value. Unless that base-level of value creation is there, all the improvements in the world won't make much difference. That makes product-market fit a precursor to product success. Though some people believe that testing is a "silver bullet"—it definitely isn't. Products don't generate exponential results simply by running lots of tests on them. Products must create real value for people first before that value can be compounded.

Once you've got product-market fit, then you can build it right. Building it "right" increases the business's investment. From a product perspective, experimentation is used as risk management to improve confidence in an idea before increasing spending to flesh it out. Many target users are very tolerant of quality issues when they find a product that solves a real problem. When that solution works without bugs and is fast—they overlook usability issues. After the bugs and performance issues are solved, then UX improvements are the next opportunity. This order isn't absolute. You can fix small bugs and UX issues on the way to product-market fit. There's also no harm in solving a big UX issue while improving performance. The boundaries should be blurred. A good way to blur those lines is by thinking about experiments as if they belong in different buckets.

Experiment Buckets, Buckets of Fun

The Four Buckets of Good Experiment Ideas to guide your prioritization approach are:

- Product and market foundations
- Bugs and performance
- Content and motivation
- Accessibility and usability

These buckets push the boundaries of who comes up with and prioritizes ideas. Every person finds different things important, be it because of different roles, education, gender, lived-experiences, or anything else. As such, every member of your team can identify different customer needs or problems. To get the most value out of your team's collective intelligence, explore each bucket equally. Engaging every person on your team—no matter if they are an engineer, writer, or whatever—leads not only to more (and potentially better) experiments, but it also improves team spirit and commitment.

Bucket 1: Product and Market Foundations

Product and market foundations cover three concepts that are key to business success:

- **What the product does**—its core functionality and features.
- **Who it's for**—the product's target market.
- **How it makes money**—the revenue generation method and business model.

Many people don't experiment with product and market foundations because they are the hardest to experiment on—at least at first. Low-maturity products usually have too few users to get reliable results in a reasonable time frame, so teams are typically guided by less reliable forms of evidence. They lean heavily on expert opinion and observational studies (see the "hierarchy of evidence" in Chapter 1, the section titled "Accurate Measurement Beats Opinion") until they build a decent user base. And that's OK! What matters at that stage is learning how to gather reliable evidence and building the capability to experiment more robustly later. Some examples of ideas in this bucket area are expanding your market by experimenting with different customer lead sources, introducing a new product or service, or adding a complex feature such as a wish list or comparison tool.

> **TIP** IT'S HARD—BUT NOT IMPOSSIBLE—TO EXPERIMENT ON PRODUCT FOUNDATIONS
>
> Experimenting with product foundations becomes easier during later phases of product maturity. But it's not impossible to do with low-maturity products. You just need to aim for a huge impact and run experiments longer to increase your sample size to detect a signal. There's a cost associated with experimenting in this phase of maturity, but there's also significant risk in just rolling stuff out and investing in the wrong features and functionality. ∎

Bucket 2: Bugs and Performance

Bugs are problems that make a product not behave as the creator intends or how the user expects. Performance is how fast, stable, and well something works. Engineers provide an important perspective because they see experiment opportunities that others can't. Some ideas in this bucket are page load, time-to-interactive (TTI) improvements, animation hitch and screen-jump reduction, and algorithm refinements.

Bucket 3: Content and Motivation

Content and motivation are what people see or read and how they connect to the tasks they need to complete, the values they hold, and their identity. Experiment examples that fit in this bucket are refining company unique selling points, providing positive feedback to keep up momentum, and providing the right type, size, and variety of product photos to help people make the right choice.

Bucket 4: Accessibility and Usability

Accessibility and usability cover how easy things are to access, understand, and use. Experiment examples that fit into this bucket are ensuring strong focus states for keyboard-only users, optimizing color palettes to adhere to accessibility guidelines, and reducing the opportunity to make an error during form-field completion.

> **TIP** ADHERE TO THE WEB CONTENT ACCESSIBILITY GUIDELINES (WCAG)
>
> These WCAG guidelines are the gold standard for how to design accessible digital experiences. You can read them at www.w3.org. The more accessible and usable your products are, the wider your potential market is. Try to ensure that your products not only adhere to these guidelines, but also that every experiment you run meets basic accessibility requirements. Check out Sarah Horton and Whitney Quesenbery's book, *A Web for Everyone*, by Rosenfeld Media to learn more about accessibility. ∎

Keep It Balanced with a 60/40 Split

One way to strike a healthy balance is to use a 60/40 split. This split occurs when 60% of the tests that get run are low-hanging fruit or "quick wins," while the other 40% are riskier, more complex ones. To find that 60/40 split, use the Impact Effort matrix illustrated in Figure 4.1.

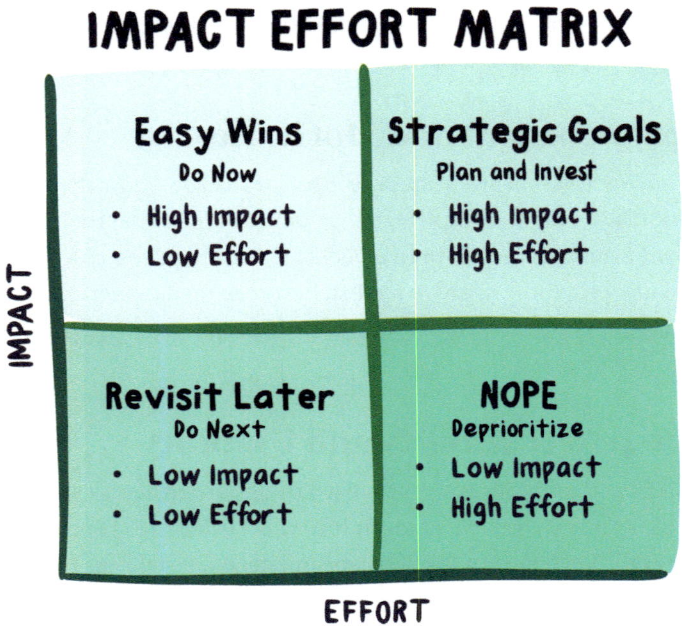

IMPACT EFFORT MATRIX

IMPACT	**Easy Wins** Do Now • High Impact • Low Effort	**Strategic Goals** Plan and Invest • High Impact • High Effort
	Revisit Later Do Next • Low Impact • Low Effort	**NOPE** Deprioritize • Low Impact • High Effort

EFFORT

FIGURE 4.1 Impact Effort Matrix helps teams pick high-impact tasks and projects to work on.

Easy wins are the things that potentially deliver lots of value and are relatively easy to do. Do these first and spend about 60% of your time on them. *Strategic goals* require more effort than quick wins, but they might be of strategic business importance. They typically require planning, investment, and coordination. The *revisit later* ideas are prioritized below the easy wins and strategic goals. Sometimes, as the other ideas get done, these ideas move into one of the other categories as their relative value goes up. The *nope* zone is where low impact/high effort things go.

> **TIP** IT'S IMPOSSIBLE TO KNOW IF AN IDEA IS "HIGH VALUE" UNTIL YOU TEST IT
>
> Some people think that having a "strong conviction" that something will be valuable means that it will deliver that value. Sometimes, this is a reason that people cite so they don't "waste time" experimenting on an idea. They just want to "roll it out" to get the assumed benefits of the idea. However, I've seen many "high conviction" ideas cost businesses more money than they're worth and eventually be rolled back. We test because we cannot know the true value of an idea and its execution without doing so. ■

Once the matrix is drawn, have a conversation with your team about effort vs. impact. Plot and compare the ideas against one another and shift them around until the comparison looks accurate. Finally, balance the 60% easy wins with the 40% strategic goals.

> **TIP** FOCUS ON LEARNING FAST
>
> Running many small tests helps you learn quickly what customers do and do not care about. Successful tests become a signal for which direction you should take to further improve your product as you begin to understand what customers value. Then your strategic goals can be shaped by what's proven to be important to customers, increasing your bigger bets' chances of success. ■

Working this way enables regular returns while investing in strategic improvements. This approach minimizes your chances of getting stuck in a local maximum.

LOCAL AND GLOBAL MAXIMUM

When you start testing, someone at some point will eventually say to you, "BUT WHAT ABOUT LOCAL MAXIMA?" A local maximum in mathematics is a point in a dataset that's higher than all the surrounding ones. A global maximum is the highest value in a data set as shown in Figure 4.2.

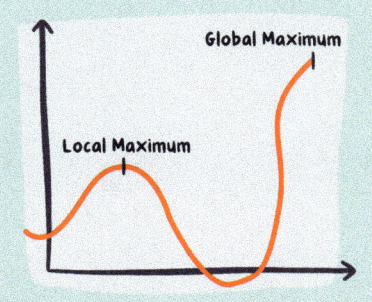

FIGURE 4.2 This illustrates the concepts of a local and global maximum in a data set.

CONTINUES ➤

CONTINUED ▶

To understand these terms, think of a mountain range such as the Alps. Many mountain peaks in the Alps tower high above the surrounding land, including Dufourspitze.[1] Once a climber makes it to the top of Mt. Dufourspitze, they can't go any higher in that area—they've hit a local maximum. Similarly, a design can be optimized until it can't possibly get any better. When a design is "as good as it can get," it's at a local maximum. You might also hear people speak about this in terms of "diminishing returns"—where each improvement you make has less and less of an impact because there's not much more to be gained from investing more.

The only way out of a local maximum is to find another peak to climb, or, in design terms—do a redesign. Eventually, through exploration, climbers find the highest peak in the Alps, Mont Blanc. That's the global maximum in the Alps. There is no higher peak available in that mountain range. Well-guided assumptions and skillful execution enable companies to make that leap. By the time a product is optimized to its local maximum, whoever has done that work has gained a lot of wisdom along the way. The likelihood they've got the skill to jump to the next mountain is way better than someone who didn't travel that path. This enables them to learn so they can reach the next highest design peak—a better base.

1. Why, yes—I did have to Google that. And no, I don't actually know anything about mountains.

Although the tech world is full of roadmaps, backlogs, projects, and programs, these approaches take a lot of time and effort. Other ways to work—some of which throw fancy prioritization right out the window—are way more efficient and impactful.

Don't Lose Sight of What Matters

Experimentation programs that lean too heavily on easy wins can become short-sighted. This happens when teams only focus on small changes or the most valuable customer base—to the detriment of long-term strategic success and diversifying their customer base.

To illustrate this problem, imagine you work at a company that gets 80% of its revenue from 20% of its users. A real-life example is a tool like Figma, the design program. Many people use it for free. A freemium pricing structure drives unpaid personal use because typically people don't want to spend money on their side projects. The unpaid personal use drives paid enterprise adoption.

If an experiment backlog is filled according to short-term sales impact, you might focus all your time improving things for the 20% of paid users. Eventually, as was the case with Sketch (a Figma competitor), unpaid user needs go unmet. When that happens, it becomes easier for someone to swoop in and shift that unpaid user onto their platform. Then the paying enterprise adopts the competing tool their employees now prefer. Sure enough, in time, the enterprise downsizes its paid licenses for the old tool in favor of the competing tool. This is exactly what happened with the more than 300-person strong design team at Booking.com. Sketch clearly optimized for its paid user base while Figma spent more time creating a great free experience. After a few years of supporting both, the whole design team switched from Sketch to Figma. This kind of strategic loss comes slowly—then all at once.

RUN A/A TESTS TO GET THE "LAY OF THE LAND"

An A/A test, as shown in Figure 4.3, is similar to an A/B test.

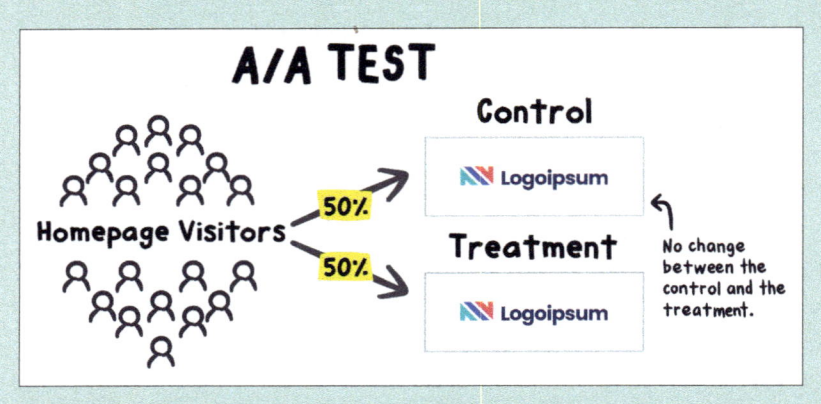

FIGURE 4.3 In an A/A test, there is no change between base and variant. A/A tests check your experiment setup.

Instead of introducing a change to half the traffic, you show no change to anyone. Doing this helps you better understand the specific area where you plan to run your experiment. In some cases, such as a complex test setup with on-view or on-click tracking, it's helpful to know if you're likely to observe a sample ratio mismatch (SRM), or if the traffic is so low you're unlikely to get the sample size you need within a reasonable time frame. A/As are literally a test run on the real product terrain so you can learn exactly where it makes sense to spend your time and if your experiment data and setup are correct. A/A tests also safely reveal any unanticipated biases or factors in the experiment design that would impact the validity of an upcoming A/B test run in those same conditions.

Find Your (Work)Flow

Everything can be optimized—including workflows. The workflow you choose impacts the speed at which you can learn and deliver value-creating improvements to customers.[2] A workflow encompasses the process, steps, and tasks needed to do value-delivering work. Effectiveness, efficiency, and productivity are all key concepts in that success.

- **Effectiveness** is how well the completed work meets the goal of creating value. The point of most prioritization is to increase effectiveness. Many companies strive for excellent prioritization, so employees only work on things that deliver direct value.
- **Efficiency** is how well resources (employees, tools, money, materials, etc.) are used to do the task.
- **Productivity** is how many tasks workers can deliver in a given time frame.

Processes can be efficient but not effective, because even the fanciest spreadsheet in the world and the strongest convictions can't guarantee high-impact priorities. Workers can be productive but not efficient. And tasks can be effective but not efficient by plowing through more resources than necessary. Tried-and-true workflow design methods strike a good balance between effectiveness, efficiency, and productivity.

> **TIP** SPEED VS. VELOCITY
>
> Shipping new things is often viewed as progress. But progress isn't just about getting things done, it's moving forward. Many teams prioritize speed because it feels like progress. But speed is one dimensional and only considers magnitude—how fast something goes. What teams should aim for instead is velocity. Velocity considers both magnitude (speed) and direction. Many people forget that things can go very fast yet not move forward. They can even go superfast—but backward! Optimizing your workflow improves your speed in running experiments. And getting good data and writing good hypotheses will increase your velocity. Sometimes, you have to slow down to make progress quicker.[3] ■

---

2. "How the World's Best Companies Build Unbeatable Focus," Conversion Rate Experts, May 2023, https://conversion-rate-experts.com/focus/
3. Lukas Vermeer, "No Velocity Without Speed," February 2, 2023, www.lukasvermeer.nl/publications/2023/02/02/no-velocity-without-speed.html#:~:text=Speed%20is%20not%20velocity%2C%20but,too%20obvious%20to%20even%20mention

Goal with the Flow

The goal of most workflow design is to deliver effective value creation to customers at maximum efficiency. That means that tasks don't sit around without getting worked on, and workers don't sit around without work. Figure 4.4 illustrates maximum efficiency in a generic workflow.

MAXIMUM EFFICIENCY WORKFLOW
Workers Always Have a Task And Every Task Is Always Being Worked On

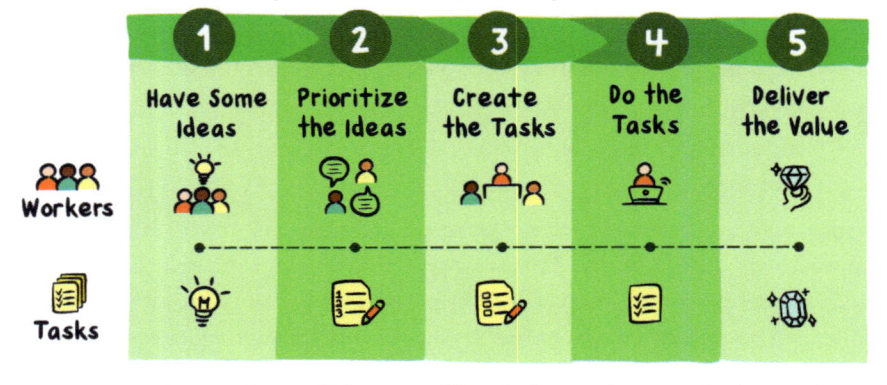

1	2	3	4	5
Have Some Ideas	Prioritize the Ideas	Create the Tasks	Do the Tasks	Deliver the Value

FIGURE 4.4 In a maximum efficiency workflow, tasks are always worked on, and workers always have a task. There is no lag time between the steps.

Workflows can be optimized for two types of efficiency: worker efficiency and task efficiency. *Worker efficiency* maximizes the tasks available so that a worker always has something to work on. *Task efficiency* is when tasks never sit around waiting—the task is always worked on. In this scenario, workers can be left idle, there's no large backlog to manage, and things get picked up "on-the-fly." At this point you might be thinking, "Gee, I wonder what kind of workflow my company has?" To find that out, ask yourself these questions:

- Are people sometimes doing "busy work" that adds no direct value to the business?
- Do people sometimes have nothing to work on?
- Do people work with short to-do lists instead of large backlog tools?

If the answer to these questions is mostly yes, then your company is probably optimized for task efficiency. If the answer is mostly no, then ask yourself these questions:

- Do people always have more work to do than time available?
- Is there a roadmap with lots of tasks and a backlog with many ideas?
- Do you have long, complex planning periods where work gets coordinated between teams?
- Do program and project managers set goals and have timelines for stuff?

If the answer to most of these questions is yes, then you probably work in a company that optimizes for worker efficiency.

The Unbeaten Path

Most companies optimize for worker efficiency. This type of efficiency leads to big backlogs and complex roadmaps where there's always more work to do than workers to do it. Optimizing for task efficiency is slightly better because it means things don't sit around waiting for someone to work on them. They keep flowing through the production line. However, there's a third option: optimize for both. Before you learn how to do that, it's good to know how things can go wrong to avoid costly problems.

Beware: "The Four Bs"

The four most common workflow problems are "The Four Bs." They are bottlenecks, blockers, breakdowns, and bloat. Each one negatively impacts the value cycle (see Chapter 1, "Conversion Design Drives Impact") in a unique way.

Bottlenecks

A bottleneck happens when tasks don't get done at the same pace. Some tasks get done fast, and others slow. That causes tasks to pile up on either side of the worker illustrated in Figure 4.5.

BOTTLENECKED WORKFLOW
Tasks Don't Get Done At the Same Pace, And Work Piles Up

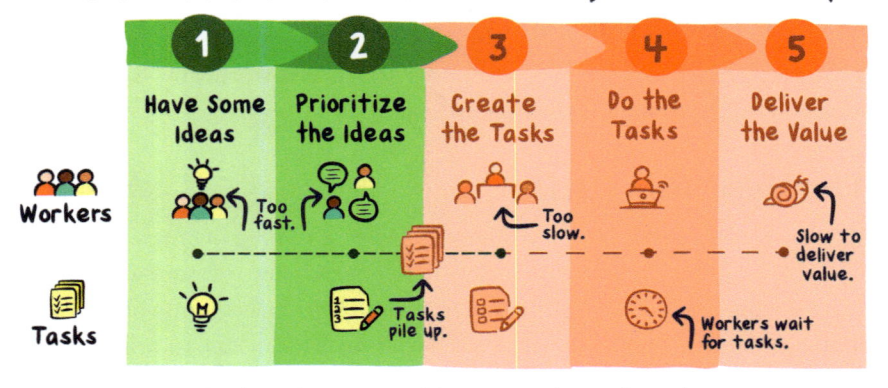

	1 Have Some Ideas	2 Prioritize the Ideas	3 Create the Tasks	4 Do the Tasks	5 Deliver the Value
Workers		Too fast.	Too slow.		Slow to deliver value.
Tasks			Tasks pile up.	Workers wait for tasks.	

FIGURE 4.5 In a bottleneck, one part of the process doesn't have the capacity to finish its task at the right pace. This causes tasks to pile up, and some workers are left without tasks.

This bottleneck means that somewhere along the line workers have to wait for tasks. Bottlenecks reduce worker efficiency and happen for a number of reasons.

- **Understaffing.** This happens when there are more tasks than workers to do them at a healthy pace.
- **Inconsistent availability.** People aren't 100% available all the time. Vacations and sickness cause bottlenecks if a task gets delayed while an important worker is unavailable.
- **Worker skill and speed.** The level of skill and capacity to complete work varies person-to-person. As companies scale quickly, they often have more new workers than experienced ones. It takes 6–12 months before a worker reaches their performance potential. New hires literally need time to "get up-to-speed." While they do, bottlenecks are common.

No process can move faster than its slowest stage—the bottleneck—and that one bottleneck constrains the entire process. So, once you've identified the bottleneck, it's irrational to work on anything else until you've fixed it.[4]

Blockers

Blockers happen when tasks can't be done. No progress can be made until the blocker gets worked around or is fixed. Blockers happen for various reasons. Not having access to important information is one. Information is anything from credentials to log in to a system, the name of a contact at a supplier, or historical data on which to build a forecast. Delays in decision-making also block work—for example, budget approvals and reviews before purchasing important supplies, or choosing which teammate to hire to fill an important skill gap. And finally, another big reason for blockers is lack of access to appropriate tools. These are things like broken laptops, systems that go down, or tools that lack critical functionality to get a task across the finish line. Blockers, like bottlenecks, cause work to build up around workers. So, again, some workers are left without tasks, while others have too many, which causes both task and worker inefficiency—not to mention frustration (Figure 4.6).

BLOCKED WORKFLOW
All Progress Stops Because Something Is Wrong or Missing

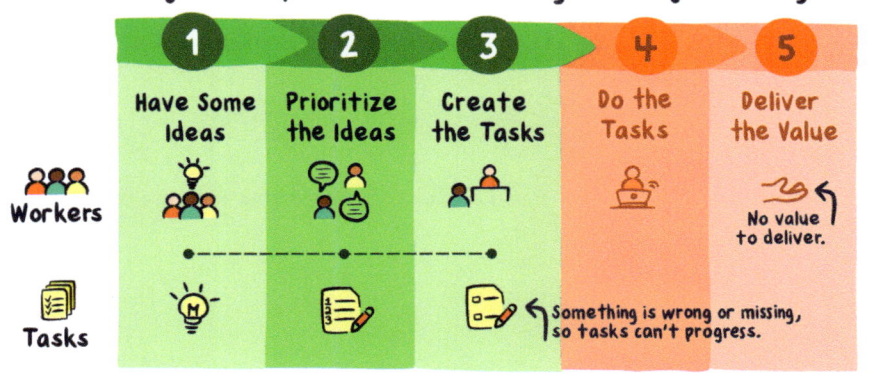

FIGURE 4.6 Blockers cause tasks to pile up when another task can't be completed for some reason.

---

4. "The Theory of Constraints: How We Relentlessly Improve Our Productivity," Conversion Rate Experts, September 2023, https://conversion-rate-experts.com/theory-of-constraints/

Breakdowns

Breakdowns occur when something in the process falls apart, which causes value to get lost along the way. Poor or one-sided communication is a big cause of breakdowns, and it causes tasks to be misunderstood, work to get lost, or tasks to be forgotten.

Quality also degrades during a breakdown. People "cut corners" to keep pace, or workers don't have the right skills to meet the standard. The value potential of the idea can also degrade as time passes. Some ideas rely on seasonal delivery to maximize their impact. Delivery that comprises a new toy line, for example, will be most impactful right before the holiday season. If the window of opportunity to maximize value passes, this is a breakdown (Figure 4.7).

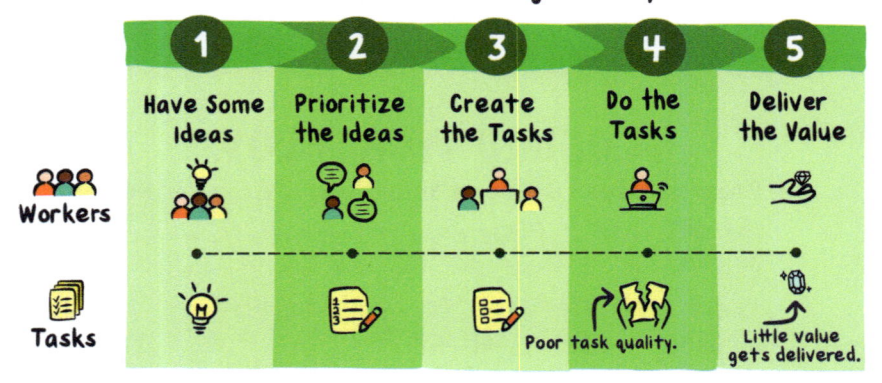

FIGURE 4.7 When a process breaks down, it reduces the potential value of the idea.

Bloat

Bloat happens when secondary needs are created to manage the process and tasks. A *secondary need* is work that provides no direct value to customers but supports the value-creating work. You'll notice bloat when task management activities take up more and more time. You might even hire a project or program manager to coordinate, document, and track work. Recurring meetings are often scheduled to minimize miscommunication during task handoffs and to manage stakeholders. But if there are too many, that are too long, with too many people—it's bloat. Finally, lengthy approval processes also indicate organizational bloat (Figure 4.8).

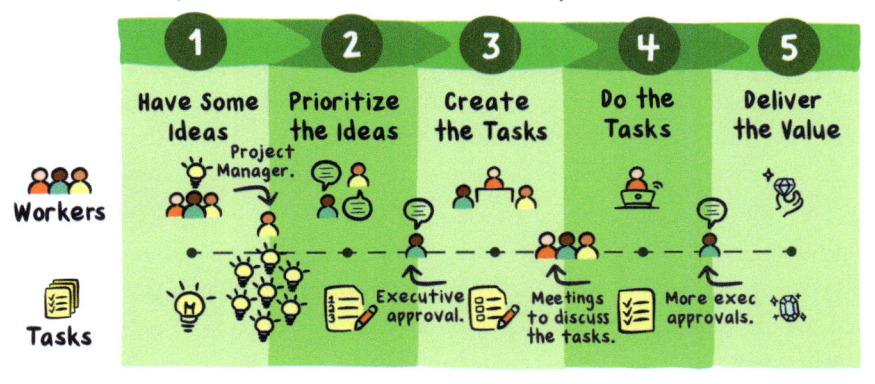

FIGURE 4.8 A bloated process creates secondary needs that provide no direct value to customers.

Kanban is a tool to help you keep the Four Bs from creeping into your workflow.

Can You Even Kanban?

Back in the 1940s, a Japanese engineer named Taiichi Ohno invented a manufacturing process called *Kanban*. *Kanban* is a Japanese word that means *visual signal* or *card*. The Kanban system relies on visual signals, such as cards, to indicate when materials are needed in the production process. To create this concept, Ohno was tasked by Toyota, a Japanese car maker, with optimizing their production process. The goal was to reduce waste and increase efficiency. To do this, Ohno drew inspiration from an unlikely source: a grocery store. Grocery stores always seem to keep the right number of products on shelves—not too much to cause waste, and not too little to cause shortages.

Throw It in the Bin

Grocery stores use a "three-bin" stocking process. The first bin is the supplier that provides the products to the stores. The second bin is the local warehouses and back stockrooms that the store shelves get replenished with. The third "bin" is the store shelf where customers choose their products. The three-bin system is based on the concept of "pulling" instead of "pushing."

The push-and-pull process deals with how work flows through a system. You can apply the same thinking to products on a shelf. *Pushing* is when the amount of stock gets based on a forecast or an estimate of a product's demand—in other words—you order an amount to sell based on a guess. When pushing, goods are purchased in large batches. This approach aims to maximize sales efficiency and minimize cost.

Pulling is when the decision to order more stock is based on a restocking threshold. For example, when a product has only two units on the store shelf, that signals when a worker should grab some from back stock and place them on the shelf. The same principle applies to items held in the stock room. When back stock hits a certain number of products, it signals that an order should be placed from the supplier. The situation determines what's done vs. an estimate or forecast determining what's done.

Kanban in Conversion Design

A Kanban board (Figure 4.9) is a visual way for teams to track work by focusing on continuous delivery.

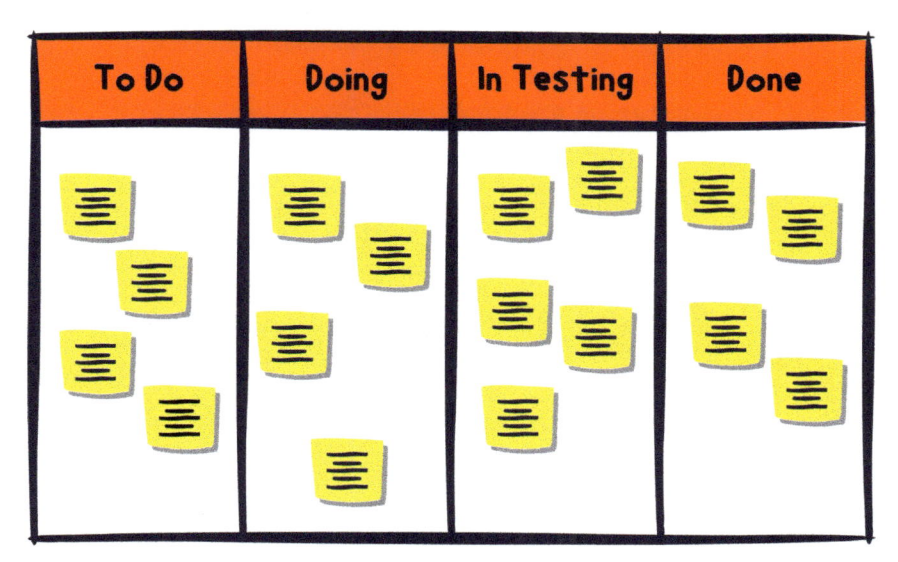

FIGURE 4.9 A Kanban board is streamlined and finite. You can only add so much to it.

Instead of producing goods or stocking shelves, teams complete tasks. In experimentation, they design, develop, and run tests. Instead of planning an experimentation program and roadmap (estimates and forecasting), Kanban lets the situation determine what gets done. The idea behind Kanban is to show all tasks with their status in a single view. By nature, this reduces the amount of backlog tasks and works-in-progress because of the finite amount of space allotted to each bin. Once you run out of space in a bin, you can't add more tasks. This approach optimizes for task efficiency.

> **TIP** DITCH YOUR FANCY BACKLOG TOOL
>
> Fancy backlog tools enable you to tag, group, sort, and search through tasks. This robust task management functionality defeats the entire purpose of explicitly limiting the number of items that go into your board to keep complexity low. Resist the temptation to record and organize every idea you and your team have. ∎

EXPERT ADVICE FROM THE FIELD

HOW TO GET MORE DONE WITH MARTIN STONE

The most successful businesses I've worked with share a singular quality—focus. They get overwhelming amounts of work done. Not just any work, but the key projects that will unlock growth. To achieve this focus, you need to follow two simple rules.

1. Keep your work-in-progress low.

    Don't let the rate of projects you start exceed the rate of projects you complete. Otherwise, chaos ensues. As work builds up, confusion, blockages, and secondary work dominate. You'll start projects to manage the projects, and progress will stall.

    So, once you've identified what to work on, make that project royalty. Give it the resources it needs. Remove distractions and guard against diversions, scope creep, and feature bloat.

    Then finish what's in front of you before starting something new.

2. Break the constraints in your projects.

    Every project is constrained by its slowest stage—the bottleneck or constraint. So, it follows that the only way to get more done is to focus on fixing—or breaking—this constraint. Working on anything else is irrational.

    Once you've fixed that constraint, it will reveal another one because there's always something slowing you down. Be relentless and focus on the next constraint and the one after that. You'll be amazed at how much you can get done.

Martin Stone
CEO of Conversion Rate Experts

Martin is a former aerospace engineer and aerodynamicist, Martin realized that the techniques he used to optimize state-of-the-art aircraft could be applied to optimizing web businesses.

Tips to Design an Efficient and Effective Workflow

If you keep your eyes open for the Four Bs and use these workflow optimization techniques, you'll learn quickly and move your value cycle forward fast:

- **Reduce the number of handoffs.** Handoffs happen when responsibility, knowledge, action, or feedback are split between two or more people. The fewer people needed to do the tasks, the less time is wasted communicating, documenting, managing, and improving.

- **Hire versatile people who can do most tasks quickly themselves.** This reduces handoffs. If you cannot find these generalists, then the next best thing is to hire people with complementary skill sets on the same, small team. This enables the team to work as a single unit, even if the work can't be done by a single individual.

- **Use light, frequent communication methods.** When work gets split between multiple people, you must communicate frequently—but not in large, long, recurring meetings. Meetings slow everything down and reduce worker efficiency. The goal of the communication should be to identify blockers, bloat, bottlenecks, and breakdowns fast and resolve them.

- **Minimize approval processes.** Approvals take time and can cause bottlenecks when approvers are out of the office or are too busy to get to requests quickly. Lengthy approval processes also signal lack of trust from leadership, which can cause reluctance to take risks necessary to truly innovate. Education is a better way to minimize errors without slowing things down.

- **Ship the simplest form of "what delivers value."** Do tasks in their most basic form that still delivers most of the value. This allows the tasks to generate value quicker while you decide if additional improvements are worth the investment.

- **Extract all value possible after its initial delivery.** Think about what improvements might create even more impact on top of the work you just delivered. When you find one impactful feature or idea, more impact is typically close by waiting to be discovered.
- **Continuously identify and eliminate secondary needs.** Identify secondary needs by asking yourself, "Would I willingly write a check to pay for someone else to do this?" If not, it's likely a secondary need, and you probably shouldn't do it.

The Important Bits

- Prioritization is ranking and doing ideas in an order, according to what's assumed to be most important.
- The goal of prioritization is to maximize the completion of effective work, efficiently.
- Use the Impact Effort matrix with a 60/40 split between easy wins and long-term goals.
- How you design your workflow matters just as much as how you prioritize your work.
- You can optimize workflows for task efficiency, worker efficiency, or, ideally, both.
- Use Kanban to keep your workflow fast and light.
- Beware of the Four Bs, which decrease workflow efficiency: bottlenecks, breakdowns, blockers, and bloat.
- Regularly reflect on your workflow and continuously make improvements to it.

CONVERSION DESIGN PROCESS

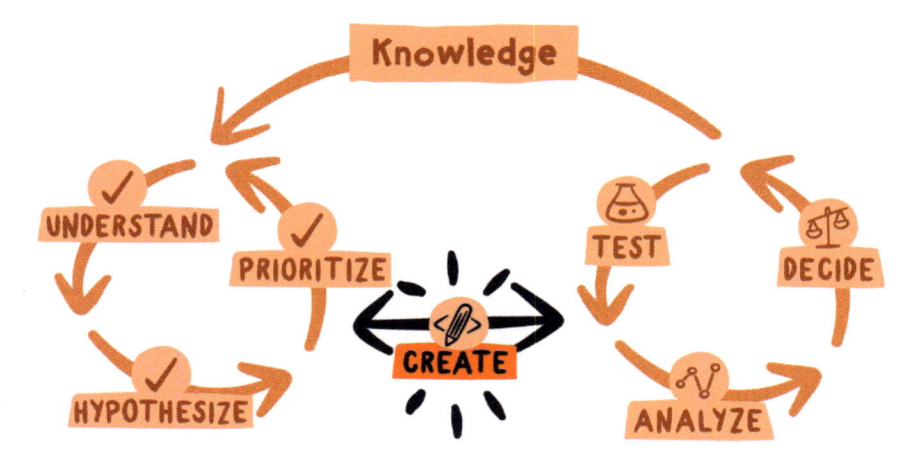

Knowledge

✓ UNDERSTAND → ✓ HYPOTHESIZE → ✓ PRIORITIZE → CREATE → TEST → ANALYZE → DECIDE

"Just enough is more."
– Milton Glaser

CHAPTER 5

The Create Phase: Set Your Idea Up for Success

THE POINT

What you design, how you design it, and how it's built all influence the impact a test can have.

I've lived about a hundred lives in my short time on this planet. In addition to working as a restaurant server and in tech, I also managed a small chain of toy stores.[1] My job was to stock the coolest toys, merchandise them in a fun way, and help customers decide what to buy. Some customers just wanted to be left alone. Others wanted personal recommendations based on a laundry list of facts about a loved one.

---

1. I have also been a Dutch-speaking flight attendant, a camp counselor, and a barista at a specialty tea shop (among other things).

In time, I got to know and care about my regular customers. I always wanted their trip to the toy store to be special. (It helped that many of them were little and quite adorable.)

Just as waiting tables helped me learn about customer experience, working at the toy store helped me learn about sales.[2] Sales at its core—whether in a physical or digital space—is about connecting with people and helping them. For example, every morning at the toy store, I'd prop the door wide open, pull the sign out, and tie balloons to the door handle. The balloons helped people find the store. And propping the door open helped my little customers and stroller-pushing parents come right in. They were small but powerful changes to the space. Though not everyone needed the door propped open or balloons to see the entrance, it made things easier for them, too. This "make things easy" mentality followed me onto the internet. And it's helped me design more effective, and broadly impactful, experiments.

To this day, I still imagine the people I'm designing for behind the screen. I imagine how I can "open the door" for them, the conversations we'd have, and how they might feel. Then, through this exercise, I imagine what a helpful human interaction could look like digitally. My aim is to design an experience that reflects the essence of that human interaction. I start with the foundations of good design: access and usability.

First Access, Then Usability, Then Aesthetics

Accessibility is if something is *possible to do*. *Usability* is if something is *easy to do*. And *aesthetic* is *how something looks*. Accessibility and usability impact everything from the words you choose to the engineering decisions you make. Most companies (including the toy store's competitors) overlooked the importance of basic access. But when people can't get in, see, and move around in your store, they either never come in, or they leave quickly. The same is true of experiments. If the experiment you design doesn't have the

---

2. Being a flight attendant taught me about how to evacuate an aircraft in under three minutes, put out fires, and do water rescues. That stuff seems less relevant to this book, though, so I left it out.

same (or better) level of access and usability as the original experience, it will likely fail. If, by accident, the access is worse in the change, you can't know whether it's the new concept you're testing that failed, or if it's a design or implementation flaw in your variant.

To set my experiments up for success, I always make sure that accessibility, usability, and aesthetics are executed flawlessly. This builds a solid design foundation for any experiment to be built on. In theory, the best approach is to just change one thing at a time. However, in practice—that's nearly impossible. This way you'll be able to tell, with a certain degree of confidence, that the outcome is due to your change, and not caused by a poor design foundation.

To execute flawlessly, I go back to my basic design principles, which are based on research. My general rule of thumb is that form (aesthetics) should always follow function (access and usability). One tool I use to frame my thinking is the Fogg Behavior Model.[3] It helps you and your team understand the elements to influence behavior change. Its creator, Dr. B. J. Fogg, provides a simple formula: B=MAP. (B stands for behavior. M is for motivation. A is for ability. And P is for prompt.) Tying it into how Conversion Design's purpose is to solve problems for people, the prompt in Dr. Fogg's model (Figure 5.1 on the next page) is the problem the customer needs to solve.

As you can see, it's easy to design experiments that are effective executions of your ideas! You just need to think about a thousand different things at once and not make any mistakes.[4] Here are some strategies to ensure flawless design and development execution on your experiments.

---

3. B. J. Fogg, Ph.D., "Fogg Behavior Model," 2007, www.behaviormodel.org
4. I'm being extremely sarcastic. The weight of all the things you need to consider can feel crushing at times.

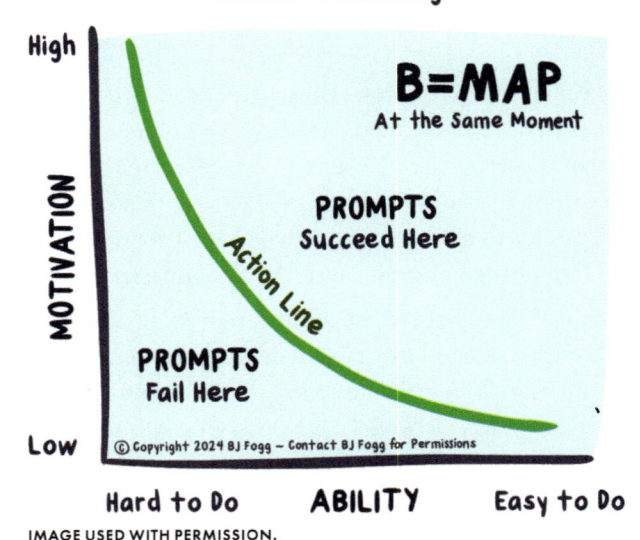

FIGURE 5.1 The Fogg Behavior Model is a framework to help analyze the design of your experiments.

Make Things Accessible and Usable

To make the information accessible to everyone, write in plain, familiar language. This approach does the following:

- **Makes reading easier for everyone.** Even if your customers are very literate, simple language saves energy and time.
- **Improves accessibility.** Simple, familiar words are more recognizable to people with dyslexia.
- **Helps non-native readers.** More than half of the people in the world are at least bilingual. That means there's a good chance that someone will use your product in a language that's not their mother tongue.
- **Aids scanning.** Many people don't read text on screens. They scan. Their brains search for "anchor words" to find what they need.
- **Is easier to scale and translate.** Software makes fewer translation errors when text is simple and clear.

You should aim to write at no higher than an eighth-grade reading level.[5] This is what's typical for a 12-to-15-year-old child. Tips to write in plain language are the following:

- **Use simple words.** Many complex words have simpler alternatives. Instead of saying "utilize," just say "use." Instead of "innovative," say "better."
- **Keep sentences short and use simple grammar.** Ninety percent of readers can understand sentences between 8–10 words long. The more words you use, the more likely you'll need complex grammar.
- **Use an active voice.** Instead of "They are aiming to achieve the goal," write "They aim to achieve the goal." You can find passive voice by looking for "ing" at the end of words.
- **Use one verb (maximum two) and front-load it in the sentence.** This builds on using an active voice, which is engaging and direct. Instead of "You should be reading, writing, and speaking often to improve your language skills," use, "Practice often to improve your language skills by reading, writing, and speaking."
- **Bold anchor words.** Because readers scan text, the quicker they find anchor words, the better. Bolding text is a great way to draw visual attention.
- **Write in short paragraphs (called *chunking*).** A chunk is a short piece of content. Chunking helps to create a visual map of where content exists in relation to other chunks. It does this by making the content look more varied.
- **Edit ruthlessly.** If there's a shorter way to say something, say it that way. If there's a word that's "nice to have," cut it. The more value-dense the content, the longer people will stay.

TIP USE THE HEMINGWAY APP

The Hemingway App, shown in Figure 5.2, is a free tool that calculates your text's reading level in real time.[6] ■

---

5. Flesch-Kincaid Readability Tests, Wikipedia, https://en.wikipedia.org/wiki/Flesch%E2%80%93Kincaid_readability_tests
6. Adam Long and Ben Long, "The Hemingway App," https://hemingwayapp.com/

FIGURE 5.2 The Hemingway App analyzes your writing to help you use plain language. It finds complex words and suggests simpler alternatives to keep you on track.

Nail Your Localization

Localization is the act of making your product *accessible* and *usable* in *different countries and cultures.* Content can be directly translated, transliterated, or transcreated. Direct translation is a word-for-word swap from one language into another. If an English headline says, "It doesn't cost an arm and a leg," other languages might not have that exact saying. Even if the words made sense and people understood the concept, it doesn't sound right. This loss of quality could make your experiment fail.

Transliteration is choosing words that keep the spirit and meaning of content familiar in other languages. In Spanish, a better option for, "It doesn't cost an arm and a leg," is "No costar un ojo de la cara." ("It doesn't cost an eye out of your face.") These sayings have similar meanings, but they're different enough to need transliteration instead of translation.

TIP OFFENDING THE FRENCH

In one insignificant experiment, a content designer tested a new email subject line: "Free breakfast deals—delicious as donuts 🍩." It showed a conclusively positive effect in North America, a weak or insignificant effect in other parts of the world, and a severely negative effect for users in France. They asked a French colleague to share insight into why the French hated it so much.

"Donuts for breakfast!? That's offensive," their colleague exclaimed. "The French eat *croissants* for breakfast," they explained. The content designer changed the content in France and French to say: "Free breakfast deals—fresh as your morning croissant 🥐." The new subject line was conclusively positive for the French, which pushed the entire experiment to have an overall positive effect. ■

Currency, number, and address format also require more than just direct translation. Some companies document local content requirements in their design systems to ensure that experiments pull from localized content foundations, as shown in Figure 5.3.

Polaris

Major currencies in their local formats

This table shows commonly-used currencies in short and explicit formats.

Currency	Locale	Short format	Explicit format
US Dollar ($, USD)	en-US	$12.50	$12.50 USD
Canadian Dollar ($, CAD)	en-CA	$12.50	$12.50 CAD
	fr-CA	12,50 $	12,50 $ CAD
Australian Dollar ($, AUD)	en-AU	$12.50	$12.50 AUD
Euro (€, EUR)	de-DE, fr-FR	12,50 €	12,50 € EUR
	en-IE	€12.50	€12.50 EUR
	nl-NL	€12,50	€12,50 EUR
British Pounds (£, GBP)	en-GB	£12.50	£12.50 GBP
Japanese Yen (¥, JPY)	ja-JP	¥1250	¥1250 JPY
New Zealand Dollar ($, NZD)	en-NZ	$12.50	$12.50 NZD
Hong Kong Dollar ($, HKD)	zh-HK	$12.50	$12.50 HKD
Singapore Dollar ($, SGD)	zh-SG	$12.50	$12.50 SGD
Danish Krone (Kr, DKK)	da-DK	12,50 kr.	12,50 kr. DKK

SOURCE: SHOPIFY'S DESIGN SYSTEM, POLARIS.

FIGURE 5.3 Number, price, and currency format are important aspects of usability.

Transcreation is when the whole experience is created from a market, linguistic, and cultural perspective. The experience feels native by getting everything just right.

Consider Cultural Differences

Localization isn't just about words and grammar. It's comprised of values, motivators, and expectations, too. What one group responds positively to might make another run in the other direction. Hofstede's Cultural Dimensions and The Culture Factor's Country Comparison Tool has six dimensions to help you understand different cultural drivers.[7] They are the following:

- **Power distance.** This measures how reliant the culture is on authority and hierarchy to maintain social order.
- **Individualism.** This captures how focused on the "self" vs. the "collective" social members are.
- **Motivation towards achievement and success.** Masculine cultures are more task-oriented, ambitious, and material-focused. Feminine cultures are more human-oriented, modest, and cooperative.
- **Uncertainty avoidance.** This captures how tolerant/intolerant a society is of ambiguity.
- **Long-term orientation.** History and tradition are important to culture with a long-term orientation. Short-term orientation seeks out modernity and progress.
- **Indulgence.** Immediate gratification and a focus on short-term gain are the hallmarks of an indulgent society. Stability and restraint are valued in less indulgent cultures.

The cultural dimensions shown in Figure 5.4 are a useful lens through which to write meaningful hypotheses and analyze experiment results.

---

7. Geert Hofstede, Ph.D., "Country Comparison Tool," The Culture Factor Group, October 26, 2023, www.hofstede-insights.com/country-comparison-tool

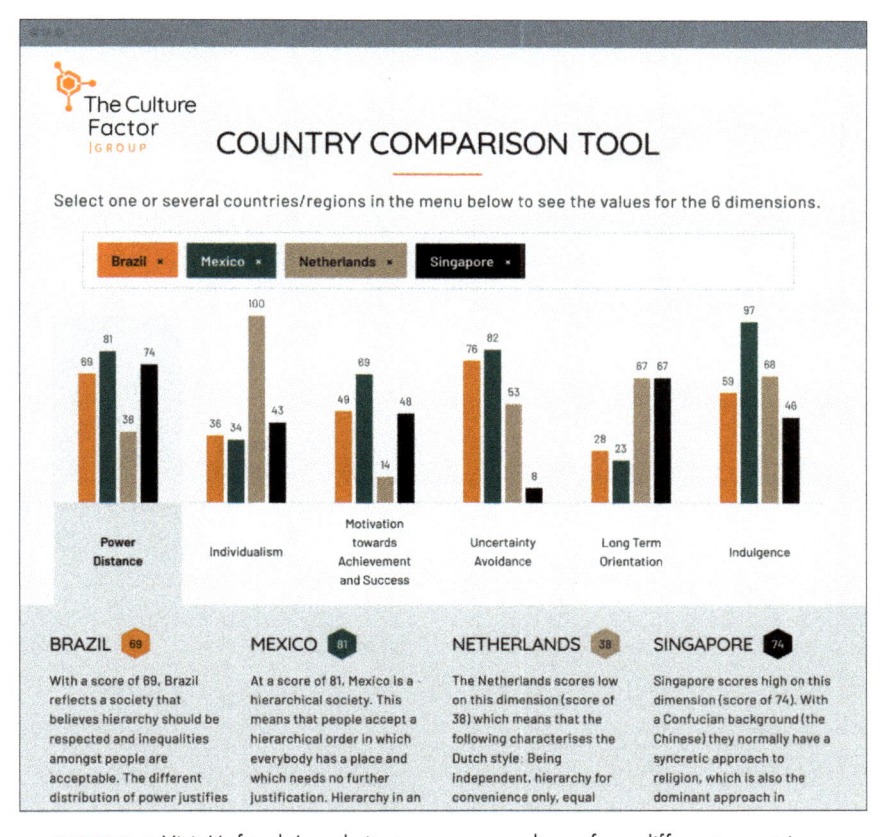

FIGURE 5.4 Visit Hofstede's website to compare cultures from different countries by Hofstede's six cultural dimensions.

For example, a highly individualistic culture might respond positively to a message about exclusivity and limited availability. On the flip side, a collectivist culture might respond negatively. When you observe different behaviors from different user groups in the data, Hofstede's model can help you hypothesize about what's happening. Then you can design your message according to the cultural drivers. These localization tweaks can sometimes be enough to make an otherwise unsuccessful experiment cross the line into significance.

EXPERT ADVICE FROM THE FIELD

CRAFTING CONTENT WITH DATA WITH FRANCESCA CATANUSO

Language choice can seem subjective, but there are ways to make data-informed decisions to ground your writing. Google Trends (trends.google.com) is a great tool to explore what people are searching for right now. You can compare between two or more words or phrases and enter parameters such as country, search dates, and topic. The output is a data visualization of things like absolute number of searches, interest over time, and related queries (Figure 5.5).

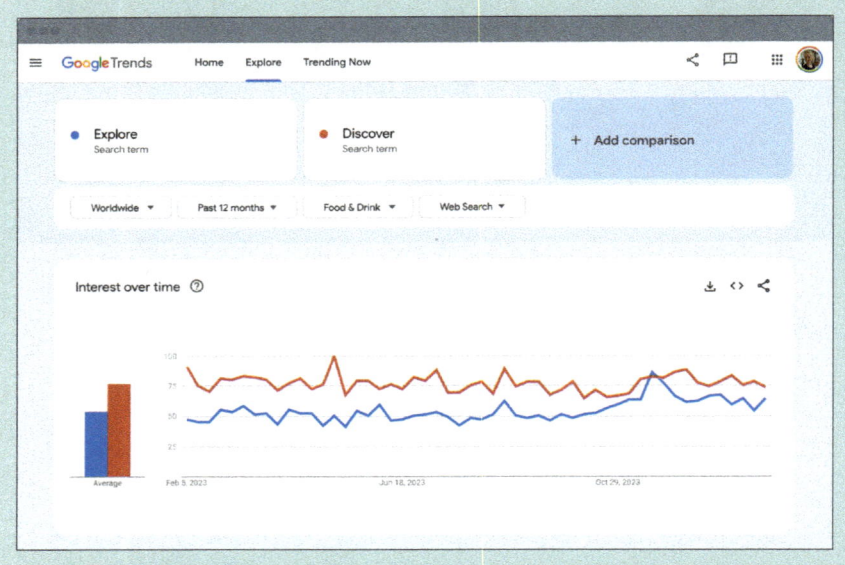

FIGURE 5.5 Google Trends data visualization comparing popularity of two words.

This information can be used in the content itself, or as part of the narrative for product teams within a content strategy brief. What's more, as you experiment to match familiar language, you can increase your conversion rates and reduce marketing spend by improving your organic search engine optimization (SEO).

Francesca Catanuso
Senior UX copywriter

Francesca, from Washington, D.C., currently based in Amsterdam, NL. She's been crafting words using data and research for more than 10 years at companies such as Booking.com, Spotify, and Meta. Outside of work, she drinks a lot of chai, and is a special-needs mom.

Write Descriptive Links and Buttons

Once you manage to entice people the whole way through your content, try to get them to take the next step with descriptive links and calls-to-action. Links that say "click here" or "read more" are disappointingly common. They provide no context about where the link takes you. Links and calls-to-action are visual "anchors" that draw the eye in. Choose the words you use in them thoughtfully. Longer, more descriptive links increase the interaction target, which makes them easier to click. This is good not only for people with fine motor issues (i.e., a tremor in their hands or arthritis), but also for search engine optimization (SEO) and screen readers (these vocalize content for Blind or low-vision users).

For example, "read more" isn't motivating or clear. The experiment in Figure 5.6 shows a help section that reduced customer service inbound calls purely by exposing all the content and making links longer and more descriptive.

Control

Quick answers to common customer questions

When will my order arrive?
Orders typically take 3–5 business days to arrive within the continental USA. Visit the Postal Service **website** and enter...
Read more

What's the return policy?
Customers have 30 days from the date of delivery to submit a return. Go to your **account** page and look under your...
Read more

Is there a warranty?
All electronics are covered by the standard 1-year **warranty**. Additional coverage can be purchased for up to...
Read more

Treatment

Quick answers to common customer questions

When will my order arrive?
Orders typically take 3–5 business days to arrive within the continental USA. **Visit the Postal Service website** and enter the tracking number. Your tracking number can be found in your shipment information email.

What's the return policy?
Customers have 30 days from the date of delivery to submit a return. **Go to your account** page and **look at your orders.** Here you can see which items are eligible for return. If an item is eligible for return, there will be a link that says, "Process return." Click that link to start **the return process.**

Is there a warranty?
All electronics are covered by the **standard 1-year warranty**. Additional coverage can be purchased for up to 5 years through our partnership with Advantage Insurance Group (AIG). **Learn more about the extended warranty from AIG on their website.**

FIGURE 5.6 "Read more" in accordion UI isn't descriptive enough to be useful to screen reader users. Additionally, these links often hide content that people want or need in FAQs or help sections.

Don't Mess with Defaults

Two things to keep your eye on are user interface (UI) states and text highlight color. Focus states (the highlight color around form fields when they've been clicked into) are designed to be highly visible. Now with dark mode becoming more common, they're also designed to work on both light and dark backgrounds. Many people who make custom focus states simply pick a brand color and don't consider how it works in different contexts. This can lead to usability issues that can impact your experiment results.

A technical experiment idea is that if you use custom cursor focus states, try replacing them with the system or browser default. One successful experiment that increased sales replaced a customer focus state with a more visible default focus state. Not only was it easier for people to see, but it also aligned more with their expectations than the custom one. Figure 5.7 illustrates the test.

FIGURE 5.7 Focus states and cursor colors are small details that make a big impact.

Focus states are visual indicators that let users know what element (if any) is actively selected. This makes focus states especially important for sighted keyboard-only users. Figure 5.8 shows what they look like in both dark and light mode.

FIGURE 5.8 The blue-and-white lines around the interactive elements are focus states. This example is of the default focus state of the web browser, Google Chrome.

For whatever reason, sometimes people add, customize, or remove focus states from HTML elements (form fields, buttons, links, etc.). They usually do it because of personal preference or for brand consistency. But messing with defaults is risky business. Defaults are how HTML tags behave in a browser when there's no CSS or JavaScript that changes how they work or look. Companies like Apple, Google, and Mozilla hire the world's best designers and researchers to create easy-to-use default interactions. Not only are they well-designed, but many people are used to and expect them to behave in a certain way. Default HTML tag behavior is a mental model that many people rely on.

> **TIP** REMEMBER JAKOB'S LAW
>
> Dr. Jakob Nielsen, a prominent usability researcher, noticed the importance of common mental models when he coined "Jakob's Law" back in the year 2000.[8] This law states that "users spend most of their time on other sites." The impact this has on your experiment design is that people form expectations of how your digital thing should work based on their experience of how other digital things work. Try not to deviate too far from what users expect in your experiment designs, unless pushing the boundary of that norm is part of your hypothesis. ∎

Text-highlight defaults, like focus state and cursor color defaults, are also commonly changed. Their primary use is to show what text is selected, which helps people to copy/paste text and images. A secondary use is to aid on-screen reading. On desktop computers, many people highlight words as they read to keep their place on the screen. When the highlight is a different color than expected, it can be jarring. Even worse—if the new highlight color is inaccessible, it can make text *harder* to read.

One negative experiment showed a significant loss in sales due to a change in highlight color. The hypothesis was that the user interface (UI) should reinforce the brand colors for consistency with the rest of the page. The prediction was that if the color used in the text highlight was the primary brand color, it would be more cohesive. The color change to the text-highlight was the only change in the experiment, so the cause/effect relationship was undeniable. Figure 5.9 shows a generic version of that test, which consistently and sharply decreased sales.

---

8. Jakob Nielsen, Phd.D., "Jakob's Law of Internet User Experience," Nielsen Norman Group, www.nngroup.com/videos/jakobs-law-internet-ux/#:~:text=summary%3A%20 users%20spend%20most%20of,for%20which%20users%20are%20accustomed

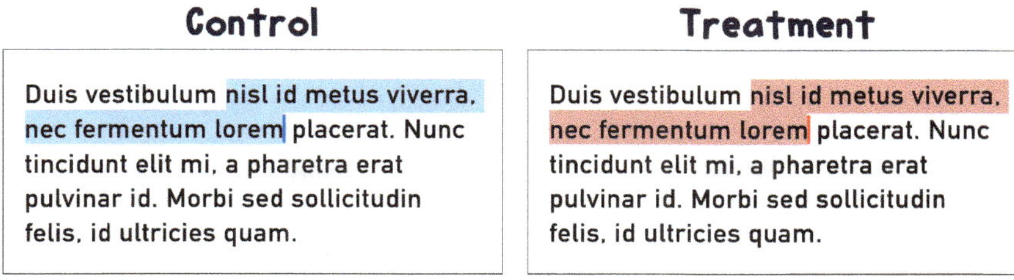

Control

Duis vestibulum nisl id metus viverra, nec fermentum lorem placerat. Nunc tincidunt elit mi, a pharetra erat pulvinar id. Morbi sed sollicitudin felis, id ultricies quam.

Treatment

Duis vestibulum nisl id metus viverra, nec fermentum lorem placerat. Nunc tincidunt elit mi, a pharetra erat pulvinar id. Morbi sed sollicitudin felis, id ultricies quam.

FIGURE 5.9 The original (left) shows the default text highlight. The change (right) shows a failed attempt to make the text-highlight the same as the brand color.

When you change a tool that people use regularly to support task completion, it needs to be a significant and clear improvement.

TIP REGULARLY VISIT THE NN/G WEBSITE FOR USABILITY RESEARCH

NN/g (Nielsen Norman Group) is a research agency that dives deep into different aspects of human/computer interaction design. They share reports about accessibility and usability that you can use to guide your design decisions. ■

Do the Thinking for People

Steve Krug's #1 bestselling UX book, *Don't Make Me Think*, is a foundational text about how to design for web and mobile usability. It's a great resource for generating research-based hypotheses. In Chapter 9 of his book, Krug advocates for "Do-it-yourself usability testing."[9] This lightweight and consistent approach to usability testing uncovers many small points of friction. As you observe people's hesitations, learn what information they find important, and hear the questions they have, you can design low-effort solutions to drastically improve usability. Moreover, many of these small changes are likely to move your important business metrics. Successful experiment ideas don't always have to be big, flashy, or strategic. They can be simple, low-effort, and practical.

---

9. Steve Krug, *Don't Make Me Think, Revisited: A Common Sense Approach to Web and Mobile Usability*, 3rd ed. (New Jersey: New Riders, 2014), 116.

EXPERT ADVICE FROM THE FIELD

MAKING THINGS EASIER WITH JONATHAN STEPHENS

Research is my favorite way to find and fix customer problems. During a usability test on a date selector for a travel company, I noticed that people spoke in terms of "number of nights." They would count the spaces between the dates or do the math in their heads to confirm their choice. "That's interesting," I thought. "Seems like a lot of work. And it's easy for people to make mistakes." Back at my desk, I whipped up an experiment to calculate the number of nights on-the-fly. I added a single line of copy that said, "From [full date] to [full date] (x nights)," as shown in Figure 5.10.

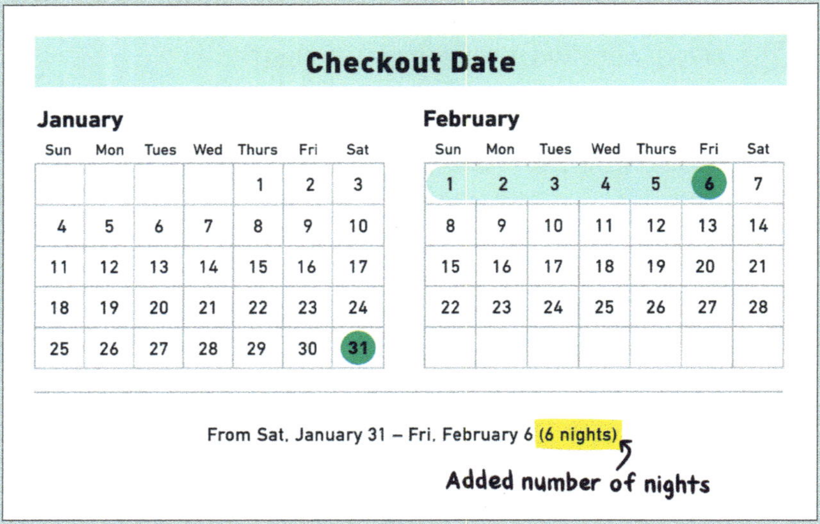

Checkout Date

January

Sun	Mon	Tues	Wed	Thurs	Fri	Sat
				1	2	3
4	5	6	7	8	9	10
11	12	13	14	15	16	17
18	19	20	21	22	23	24
25	26	27	28	29	30	**31**

February

Sun	Mon	Tues	Wed	Thurs	Fri	Sat
1	2	3	4	5	**6**	7
8	9	10	11	12	13	14
15	16	17	18	19	20	21
22	23	24	25	26	27	28

From Sat, January 31 – Fri, February 6 (6 nights)

Added number of nights

FIGURE 5.10 The highlighted line of content increased conversion by saving users mental energy.

My goal was to save people the mental energy of figuring it out themselves. After running the test for the full predetermined length with enough visitors for significance, the results were conclusively positive on completed sales. It also reduced customer service contacts.

I always try to replicate success to get as much value from an idea as possible. Once the dates were selected, I reinforced the number of nights for reassurance in the search box itself. Then I made sure that it was featured on every search box in the product. The concept continuously tested positive, which taught me a few different things. The first thing I learned was to *really* listen. Hear the words people use and reflect them in the UI. The second thing was to be helpful. A "small gesture" like counting the number of nights for someone earns goodwill. And finally, excellent design doesn't have to be fancy. Sometimes excellence is a few well-placed words.

Jonathan Stephens
Co-founder of Poet & Scribe, a creative business consultancy
(www.poetandscribe.com)

In a past life, Jonathan worked as a senior manager of software development at Booking.com. During his time at Booking, he ran hundreds of A/B tests, which elegantly solved customer problems in ways that drove business impact.

Put Things Where They Make Sense

It's a common pattern to have reviews consolidated into one area. But reviews are useful social proof that often relate to certain features or aspects of a product. You can reinforce related information by placing people's opinions near where that information is shown. For example, photo galleries, as shown in Figure 5.11, are a great place to try this.

FIGURE 5.11 In the control (left), the photo gallery had no relevant customer reviews. In the treatment (right), the photo gallery had an overlay with relevant customer reviews.

Product descriptions are another useful spot, as shown in Figure 5.12.

FIGURE 5.12 You can use generative AI to summarize customer sentiment and place it strategically near related content.

Answer Questions Along the Way

Humans have an innate need for certainty. Give them the information they need to feel in control. One way to do this is with perfectly timed nudges that address objections or thoughts as they have them. Some common objections of customers are no free shipping, a required telephone number, strict return policies, or lack of a guarantee. As you uncover these objections, run experiments to address or remove them. Deal breakers often have to do with reducing risk. An example of addressing a common objection is to state why a phone number is necessary if it's required, as shown in Figure 5.13.

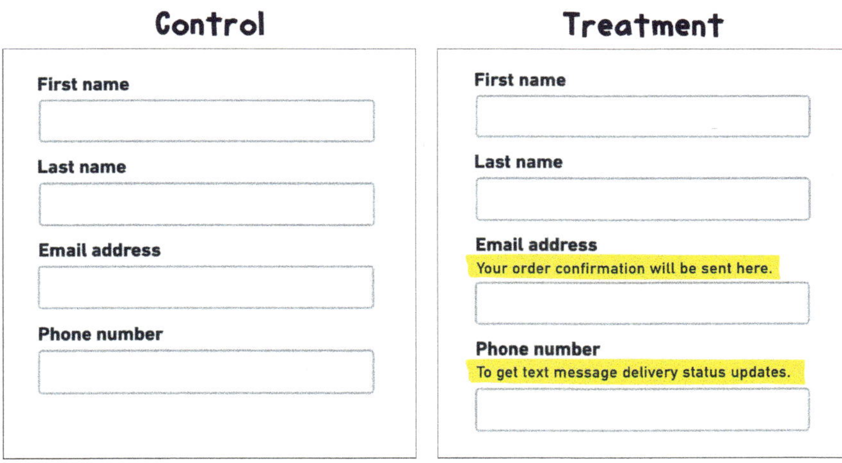

FIGURE 5.13 In the control (left), the form fields had no supporting text. In the treatment (right), supporting text explained why the information was needed.

TIP EXPERIMENT WITH TOOLTIPS

Tooltips increase the amount of useful information for those who need it (without distracting those who don't). They get placed in close proximity to something that people are looking at or are doing. Use them for helper text, reassurance, or extra information that's useful to some, but not most, people. Heatmaps can show you where people are clicking or tapping to understand where people want more information. ∎

But you don't always need to hide extra information behind a tooltip. Sometimes it makes sense to communicate things in more than one way and repeat things in different places.

Use Redundancy to Aid Understanding

A favorite quote among designers is "less is more." But what designers fail to realize is that what they believe is perfect in terms of aesthetics might be imperfect in terms of basic usability. In the words of world-renowned graphic designer, Milton Glaser, "Just enough is more." Information redundancy is when you show the same thing in more than one place or in more than one way. It helps people understand and use the information, regardless of their ability, preference, or attention span. For example, instead of just using an icon, a color, or a single word, use two or three of them together. One place where icons are useful is in a product features list. Figure 5.14 shows an example of a simple list with bullet points. By swapping the bullets with icons, you add a rich layer of information without taking up any extra space.

Control

Features people love
- Free repairs for life
- 360-degree maneuverability
- Lightweight—less than 2 lbs!
- Eco-friendly materials
- TSA-approved combination lock
- Reinforced zippers for added strength
- Extra large storage pockets for organizing

Treatment

Features people love
- Free repairs for life
- 360-degree maneuverability
- Lightweight—less than 2 lbs!
- Eco-friendly materials
- TSA-approved combination lock
- Reinforced zippers for added strength
- Extra large storage pockets for organizing

FIGURE 5.14 In the control (left), it's just a bulleted list of features. In the treatment (right), icons communicate meaning and reinforce the words.

Icons are also useful in navigation. Designers often prefer to use only icons or only text in nav bars. But if the icons aren't universally understood (and sometimes even if you think they are), it's good to include the word, too. Sometimes an icon helps reassure people. Navigation bars are usually slim and don't have icons. However, navigation that includes big icons and descriptive labels can outperform navigation without icons. In one test on a complex application (Figure 5.15), a traditional "slim and clean" looking navigation bar was tested against the existing "chunky and busy" navigation bar.

FIGURE 5.15 In the control (above), the navigation included large icons to accompany text. In the treatment (below), the icons were removed to take up less space and to look "cleaner."

Data showed that fewer people completed their tasks when using the traditional "slim and clean" navigation compared to the existing "chunky and busy" one. Research to understand the result showed that many non-native English speakers had to use the product in English because of the business environment. Some people navigated the product by recognizing the icons. When the designer removed the icons, they removed important redundancy that some people relied on.

> **TIP** BLUR THE WORDS TO CHECK YOUR VISUAL DESIGN'S CLARITY
>
> One way to test if your product has enough redundancy is to blur out or remove all the words in a design. Then do a 5-second test to see if people understand what they're looking at. Next time you hear someone say, "Yeah, but this is redundant." Reply with, "Yep! Redundancy is GOOD." ∎

Show and Tell (but Mostly Show)

Imagery is also a great way to add redundancy to reinforce a message, which makes pictures a powerful communication tool. One concept to try is instead of showing the product itself, show the benefit of the product. Deborah O'Malley shared a winning test on her site called GuessTheTest. Typically, product images are not shown on a model. They're shown against a plain background so that people can see the details of the product. But Melina Hess at DRIP Agency hypothesized that clothing items from the company SNOCKS displayed on models better communicated the purpose and benefit of the product. She predicted that if she replaced the plain product images with images of the product being worn, the items would be more appealing due to a clearer benefit.[10] Consequently, more people would buy the products. Figure 5.16 shows what she tested.

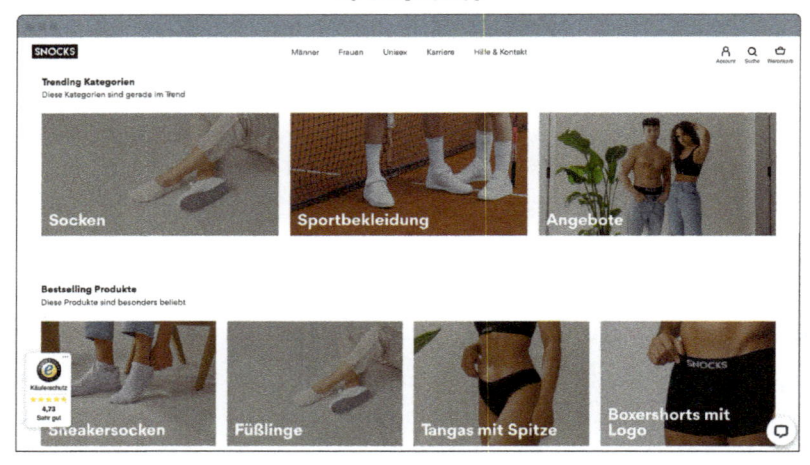

FIGURE 5.16 In the control (above), products are displayed to show detail and quality. In the treatment (below), products are displayed on models to show use and benefit.

10. Deborah O'Malley, "Guess the Test," LinkedIn, July 2023, www.linkedin.com/posts/deborahomalley_abtesting-optimization-productoptimization-activity-7085332924924837888-dPfT?utm_source=share&utm_medium=member_desktop

Images are only useful if people can access them, though. Try using image recognition software to automate the creation of rich alt-text if you have lots of photos. Alt-text is a written description of what is in the image. When you include alt-text, you make that information accessible to people who use screen readers—as well as to search engines. This can increase conversion for Blind or low-vision users, as well as improve search results ranking. A series of winning experiments used image recognition software to write and translate image alt-text at scale, and it not only improved organic SEO, but it also conclusively increased sales. This type of experiment has the potential to create free traffic that converts better!

Aesthetics Are Information

What something looks like is a powerful communication tool. Aesthetics when used strategically, make people feel a certain way, reinforce messaging, inspire people to act, or aid in task support. As such, visual design choices shouldn't impede people from achieving their goal. When you design for impact, visual design choices should enhance the experience.

> **TIP** WHEN YOU STRIKE GOLD, KEEP DIGGING
>
> A *free* label (Figure 5.17) that was added in a positive experiment was easy to miss because of its light visual weight compared to the text block next to it. So, it was knocked out on a green background to increase contrast and saturation. This change conclusively increased sales and was replicated onto all other platforms. The lessons here are when you know people care about a motivating piece of information, make it *very* easy to see and access. Then expand its impact to relevant platforms and contexts to give and get as much value from it as possible. ∎

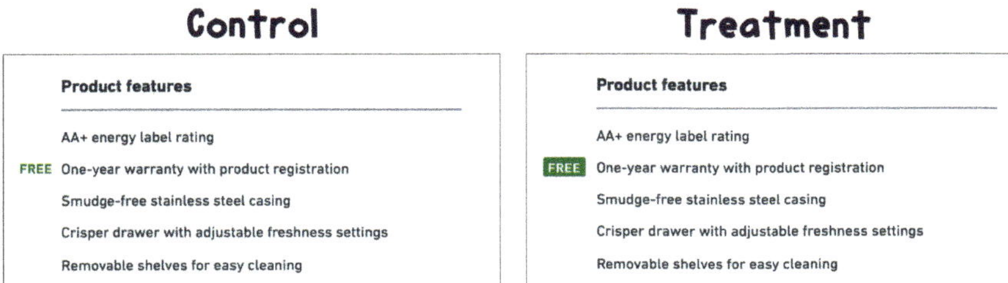

Control

Product features

AA+ energy label rating
FREE One-year warranty with product registration
Smudge-free stainless steel casing
Crisper drawer with adjustable freshness settings
Removable shelves for easy cleaning

Treatment

Product features

AA+ energy label rating
FREE One-year warranty with product registration
Smudge-free stainless steel casing
Crisper drawer with adjustable freshness settings
Removable shelves for easy cleaning

FIGURE 5.17 In the control (left), the word *free* was green on a white background. In the treatment (right), the word *free* was white text knocked out on a highly visible green.

Maintain and Fuel People's Motivation

Making something super easy, still doesn't guarantee people will do it. That's why it's important not only to focus on the "A" (ability) part of the Fogg Behavioral Model to design for impact, but you must also design for the "M" (motivation) in the B=MAP equation. Many research-based techniques exist that can help you design for motivation. As you design the execution of your experiments, think to yourself, "Is this easy and motivating?" And if it's impossible to make a complex task any simpler, then you need to lean even more on making things motivating.

Get to the Point Quickly

Even if you write clearly and concisely with perfect localization, when you don't get to the point, you lose people. Two tools to get to the point fast are the Johnson Box and the Inverted Pyramid. The Johnson Box (Figure 5.18) gives readers an overview of the content in a scannable, enticing way.[11]

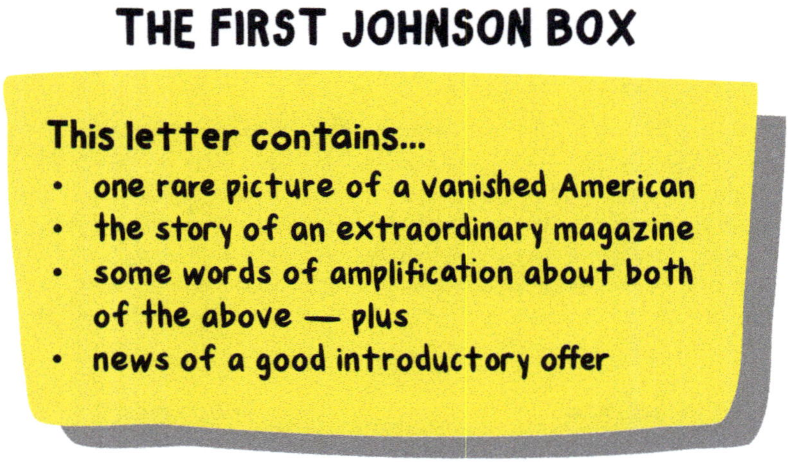

THE FIRST JOHNSON BOX

This letter contains...
- one rare picture of a vanished American
- the story of an extraordinary magazine
- some words of amplification about both of the above — plus
- news of a good introductory offer

FIGURE 5.18 Frank Johnson's first "Johnson Box" at the top of a direct mail letter for *American Heritage* Magazine.

The Johnson Box was popularized in the mid-1900s by a direct marketer named Frank Johnson. Its purpose as Johnson gruffly put it was, "To get

---

11. Chief Marketer Staff, "Frank Johnson: The Johnson Box Letter," May 1, 2001, www.chiefmarketer.com/frank-johnson-the-johnson-box-letter/

somebody to read the goddamned thing, that's all." In practice, a Johnson Box helps readers know quickly if they're interested in the content.[12] This design pattern is effective because it forces authors to write content with strong information architecture (IA) with a focus on hooking a reader. Well-written Johnson Boxes capture people's attention and allow them to quickly get to the part they find most interesting with jump-links.

The inverted pyramid (Figure 5.19) is a journalistic tool.[13] It structures content according to a reader's attention span. Use these techniques in your experiments to give your customers the information they need quickly.

THE INVERTED PYRAMID

Most Important Info
The who, what, when, where, why, and how above "the fold."

More Details
Important and unique details that are a bonus.

Conclusion
Nice to know, but not critical.

FIGURE 5.19 Put the important stuff up front and above "the fold." Yes, "the fold" still matters.

---

12. Chief Marketer Staff, "The Man Who Invented the Johnson Box," May 1, 2001, www.chiefmarketer.com/the-man-who-invented-the-johnson-box/
13. Amy Schade, "Inverted Pyramid: Writing for Comprehension," Nielsen Norman Group, February 11, 2018, www.nngroup.com/articles/inverted-pyramid/

Give People a Nudge to Get Started

If a user needs to do something to move forward, give them an example as a little nudge. When a feature is brand new, placeholder and helper text matter a *lot* to get them started. A good example is a prompt nudging people to write more detailed reviews, as shown in Figure 5.20.

FIGURE 5.20 Add prompts to forms to give people examples of what they need to do.

Give Regular and Appropriate Feedback

Feedback is dynamic information that lets people know expectations and gives information about the current situation. To encourage people to write useful reviews, set expectations about what "useful" is and let them know how they compare to that expectation. Figure 5.21 shows useful feedback to help customers write better, more helpful reviews.

FIGURE 5.21 In the control (left), the form sets no expectation and gives no encouragement. In the variant (right), dynamic feedback sets expectations and gives positive feedback.

Keep people filling in important forms with tips and visual cues (Figure 5.22). One way to do this is to use form validation when they've done something right. Form validation is a check that's done to see if the information entered follows certain rules before it's submitted. This form also allows people to correct mistakes as they go instead of learning about it after-the-fact.

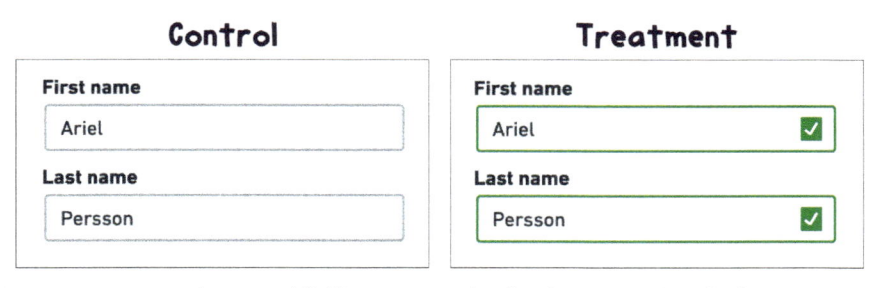

FIGURE 5.22 In the control (left), no positive feedback is given when the form is valid. In the treatment (right), positive feedback is given when it's filled out correctly.

Consistent expectation setting and in-the-moment feedback are typically the way to go. Immediate feedback should also highlight errors. Again, with on-the-fly validation, people can immediately identify if something isn't right. Then you can provide them with specific feedback on how to handle the error, so they know what to do next.

Make It Fast

You can write engaging, clear content, localize it perfectly, and entice people to the next step with a strong call-to-action, but if the page doesn't load fast, people won't stay long enough to see it. Large images, complex animation, fancy transitions, and parallax scroll can all make experiments' ideas fail. Beware when you see these things creep their way into your experiments. Animation is problematic because of these reasons:

- **Health.** Different disorders can cause dizziness or trigger seizures when viewing motion or flashing.
- **Overuse.** Animations that are too frequent or slow can block people from moving forward swiftly.
- **Weight and complexity.** Changing, removing, or adding code—especially complex code—is a risk.

Good animation is effective communication. It should answer some of these questions for people:

- Is something happening? If so, what?
- What needs my attention?
- Which direction am I going?
- How do things relate to one another?

If your experiment unintentionally improves or degrades your product's actual or perceived performance, it influences the outcome—for better or for worse. Web pages and app screens should load between 0–2 seconds. Anything above 3 seconds drastically increases the chances that visitors will bounce.[14]

> **TIP** FOR WEB PAGES, USE GOOGLE PAGESPEED INSIGHTS
>
> PageSpeed Insights report (Figure 5.23) what technical aspects of your product can be optimized to achieve conversion gains. The report lists things from image size, poorly ordered headings, and missing alt-text to help you get your code perfect in order to set your experiments up for success. ■

---

14. Google/SOASTA Research. From Think with Google, "Find Out How You Stack Up to New Industry Benchmarks for Mobile Page Speed," 2017, https://storage.googleapis.com/twg-content/original_images/mobile-page-speed-new-industry-benchmarks-01-01-download.jpg

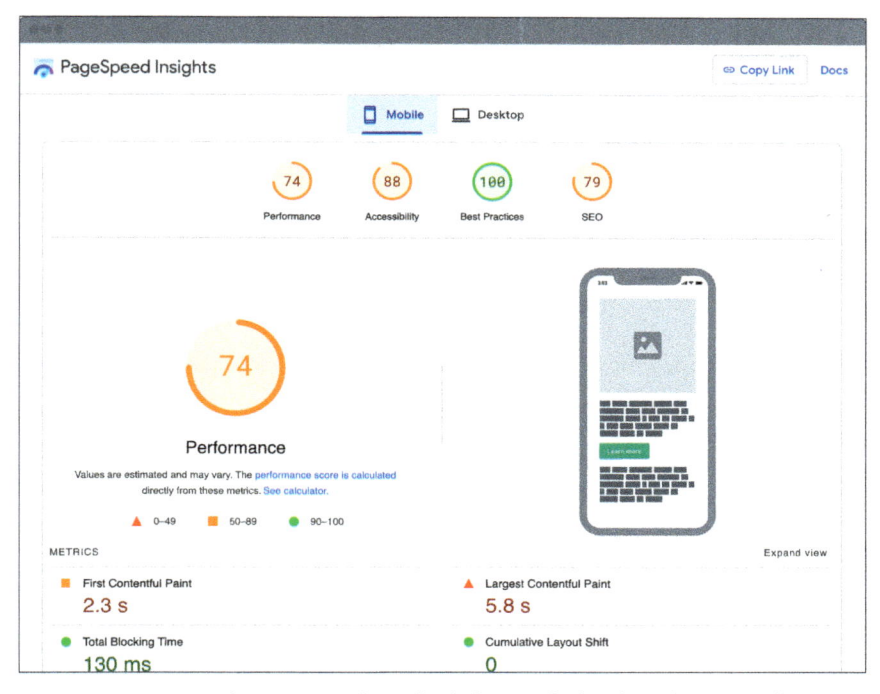

FIGURE 5.23 Google PageSpeed Insights helps you find technical aspects of your product to optimize in order to improve conversion rates.

THE CREATE PHASE: SET YOUR IDEA UP FOR SUCCESS · 119

EXPERT ADVICE FROM THE FIELD

PAGE LOAD SPEED AND DIMINISHING RETURNS WITH JORDEN LENTZ

You might be familiar with the Amazon experiment where they delayed a desktop webpage load time by 100 milliseconds.[15] The experiment decreased the company's sale by a whopping 1%. Back in 2006, around the time this experiment was run, that would have equaled a loss of about $107 million per year.

Because of this experiment, speed is one of those things that everyone knows is important. But how fast is fast enough? And is there a point at which working on more improvements leads to diminishing returns? These were questions that my team and I wanted to answer to set up performance guardrail metrics. This would help us always serve up screens in line with customer expectations. To answer these questions, we ran multiple experiments at Booking.com. The hypothesis was that app speed influences sales. But we also believed that not every 100 millisecond is the same. It's possible that delaying a product description screen will have a different impact than delaying a company info screen.

With more than 50% of Booking.com's accommodation room-nights sold coming from mobile apps, we thought it was essential to learn more about the impact of speed on our app customers. To learn this, we artificially slowed app speed and tracked the impact on both customer experience and business metrics. First, we slowed performance a lot—between 2 and 6 seconds. This proved beyond any doubt that app speed is important. Then we slowed different screens by 500 millisecond increments to learn the thresholds at which load time could increase while not severely impacting the customer experience.

---

15. Steven van Vessum, "Amazon Study: Every 100ms in Added Page Load Time Cost 1% in Revenue," *Conductor* (blog), https://www.conductor.com/academy/page-speed-resources/faq/amazon-page-speed-study/

This is what we learned:

- **First impressions matter.** Slow screens have more of a negative effect higher up in a sales funnel than lower down.
- **Migrate and modernize your technology with care.** Changing tech stacks and "modernizing" infrastructure can impact customer behavior and business metrics in unpredictable—often negative—ways.

And this is how we've changed based on what we learned:

- **We created performance dashboards.** We use these to track screen load in real time to monitor for potential issues.
- **We prioritize with performance data.** This focuses our attention on screens where performance has the most impact.
- **We set performance guardrail metrics.** These guardrails give product teams realistic and impactful service-level agreements (SLAs) to protect customer experience quality and business performance. New tech and code modernization efforts also need to be rolled out in experiments to understand any potential impact on performance.

Jorden Lentze
Senior product manager at Booking.com.

Jorden has been running experiments since 2010 and has worked for Booking.com, Google, and ABN AMRO. He lives in the beautiful city of Rotterdam in The Netherlands.

Breathe Life into Dead Ends

Dead ends are a great place to breathe life back into designs. These happen when the next step isn't immediately obvious. They come in two forms. The first type of dead end happens when the customer is following the natural flow, and it leads to nowhere. The bottom of a landing page, the last result in a search result page, and the last step in a process are some examples. In these cases, it's helpful to include another call-to-action or provide more related information the customer can move onto next.

The second type of dead end happens when people end up where they shouldn't be. These are related to errors or a product shortcoming. Some examples are 404 pages, disambiguation pages, and empty states. High-intent users sometimes end up in these places and telling them the next step gets them back on track. The best strategy to deal with errors is to make it impossible to make one. But when you find a common spot where people hit a roadblock, give them a graceful way to recover. Good error messages typically include a header that explains what went wrong, a body that says how to fix it, and a call-to-action for where to go next (check out Figure 5.24).

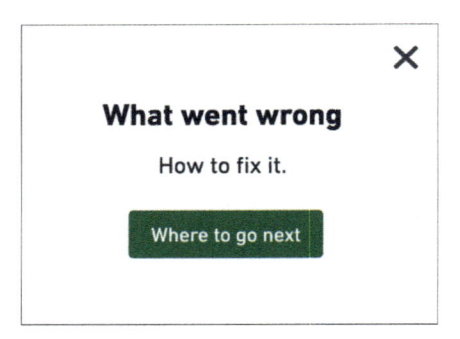

FIGURE 5.24 The three important parts of an effective error message.

Make these error messages highly visible because hidden error messages are a conversion death sentence. Whenever possible, put the error close to the problem and focus it within the screen, as shown in Figure 5.25.

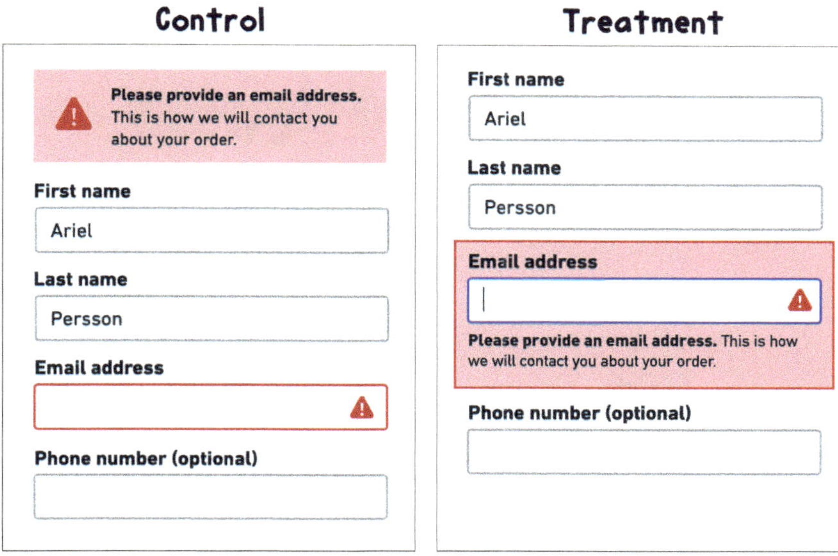

FIGURE 5.25 In the control (left), the error message doesn't visually relate to what it's referring to. In the treatment (right), the error message draws attention to what needs to be fixed, and the text field is focused.

If the error happened in a few places, scroll the user to each error one-by-one until they are all fixed. Error messages should also be clear and direct. Whenever possible, connect the heading with the call-to-action the user needs to take. Assume that people won't read the body copy and will just glance at it. Figure 5.26 shows a winning error message that improves clarity and directness.

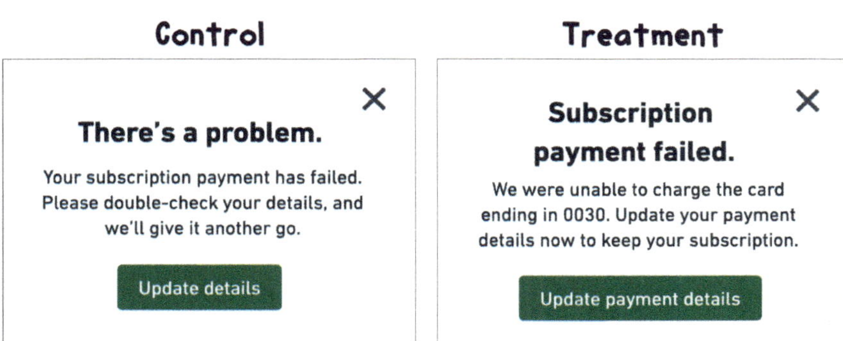

FIGURE 5.26 In the control (left), the headline is vague, and it doesn't relate directly to the call-to-action. In the treatment (right), the headline states a specific problem, and it relates to the solution in the call-to-action.

THE CREATE PHASE: SET YOUR IDEA UP FOR SUCCESS · 123

Finally, if people need to go somewhere and you can't link them directly, use carats (>) and bold text (Figure 5.27) to point the way.

Control

Unlock unlimited access

This content is available in the Gold plan. Upgrade your plan in account settings to get instant access.

Close

Treatment

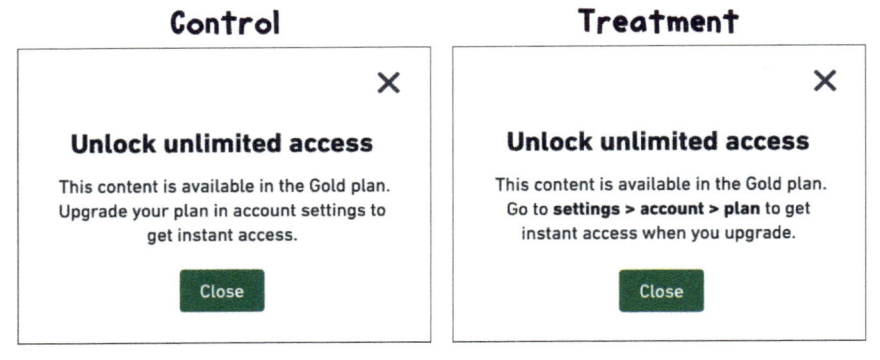

Unlock unlimited access

This content is available in the Gold plan. Go to **settings > account > plan** to get instant access when you upgrade.

Close

FIGURE 5.27 In the control (left), the instructions are written into the text. It's not immediately clear where to go. The treatment (right) uses carats and bold type to show people where they need to go.

Try to avoid these common mistakes to keep people motivated:

- **Don't alarm people.** Avoid scary words like *illegal* and *error* but do use words like *problem* or *issue*.
- **Don't be funny.** Clarity is more important than entertainment.
- **Don't focus on error codes.** If you must add an error code for engineers, make it small and briefly explain who it's for.
- **Don't blame.** Implying "blame" for an error—even subtly—kills motivation. Use a passive voice instead of an active voice in this context because it helps to dissolve blame.

EXPERT ADVICE FROM THE FIELD

KNOCKING DOWN DEAD ENDS WITH NEKEIA BOONE

A common dead end is the last picture in an image carousel or photo array. When people reach the end, typically something needs to be closed. Perhaps it's a modal or an overlay. To some, pressing the "close" button or the "back" arrow feels like just that—a step back. Instead, reiterate an important message and put a strong call-to-action to keep the forward motion. This direction guides your customer to a useful next step, as shown in Figure 5.28. It also helps people act easily on their motivation.

FIGURE 5.28 In the control, the image carousel has a dead end at the last image. In the treatment, the last image has a timed overlay with a clear next-step.

Nekeia Boone
Copywriter, brand storyteller, UX advisor, and senior tech manager

Nekeia is also the founder of Tudy's Kitchen, a Black-owned, woman-owned food brand in The Netherlands. Learn more about her at **tudyskitchen.com** *and* **nekeiaboone.com**.

Use Framing Techniques

Humans have blips in their logical processing called *cognitive biases*. When you know these biases, you can present information in a way that motivates and helps customers achieve their goals. *Framing* is how information is presented or the context in which it's introduced. It influences people's perceptions, attitudes, decisions, and behaviors. Some framing techniques to experiment with are:

- **Loss aversion.** People respond differently to gains versus losses—even if the end result is the same.
- **Scarcity.** When there's not much of something left, people's fear of missing out drives action.
- **Urgency.** When something needs to be done fast, it's urgent. Tell people how much time they have to act to shift behavior appropriately.
- **Social proof.** People look to others to guide their actions, decisions, and behavior. People feel reassured when they know others like them or people they respect made the same choice.
- **Default bias.** People usually just stick with the proposed option because it's easy. They assume that the default option is good because it's the "standard" choice.
- **Anchoring.** People are heavily influenced by the first piece of information they get (the anchor), even if it's not a good indicator of what's normal.
- **Gifting.** People want to feel special and giving them something special does that.

> **TIP** THERE'S NO SUCH THING AS A "SILVER BULLET"
>
> Just because a concept worked for a similar product with similar customers doesn't mean it will work on your product for your customers. Customer needs are different at different stages in the customer journey. In general, urgency messaging, social proof, encouragement, and nudging are useful in the upper part of a sales funnel. The post-purchase experience typically benefits from reassurance, autonomy, and control. But how and when you use them makes all the difference. To learn more about how to use psychological principles to improve your experiment execution, check out the Rosenfeld Media book, *Engaged: Designing for Behavior Change* by Amy Bucher. ■

Use Compelling Words

In the same way that framing motivates people, specific words do, too. Experiment with these influential words to learn if your customers respond to any of them:

- **The customer's name and "you."** Humans love to be seen and spoken to directly because it appeals to their egos. When you address users by their name or as "you," it engages them more than generic terms.
- **Instant.** No one likes to wait. We all love the dopamine hit of getting what we want exactly when we want it, which is instantly.
- **Because.** This is because of something called the *because justification*. (See what I did there?) When you back up a claim or request with a "because" statement, people are more likely to accept things as they are, and they become more agreeable.
- **Save.** This word invokes loss aversion. People would rather not lose something they already have vs. gain something new.
- **Easy.** A path that's framed as "easy" is usually the one that people will take. If you designed something to be easy, tell people and explain why so they believe it.

Make Tasks Manageable with Good Sequencing

Good "sequencing"—a way to break a complex task into manageable, motivating steps—makes work more efficient and rewarding. Each small action provides a dopamine hit and is a further commitment to complete the whole task. As people do one part of a task, they build positive momentum, which makes them more likely to finish. Many progress indicators start at the first step in the checkout or sign-up process that includes filling out a form. You can give people a boost of motivation by showing them they've already taken the first step, as shown in Figure 5.29.

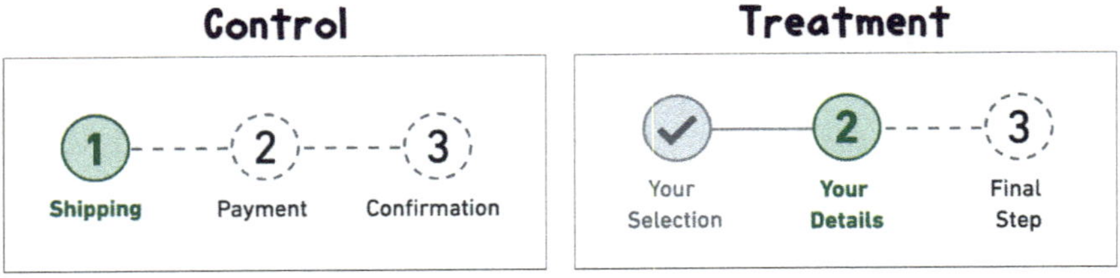

FIGURE 5.29 In the control (left), the progress indicator started on step 1. In the treatment (right), the progress indicator accounts for the work already done and starts on step 2.

When you acknowledge the work your customers have done already, you keep the momentum going for them.

Add Encouragement Along the Way

As people work their way through a difficult set of form fields, let them know they've done well! Short lines communicating their good work go a long way toward keeping them going, as shown in Figure 5.30.

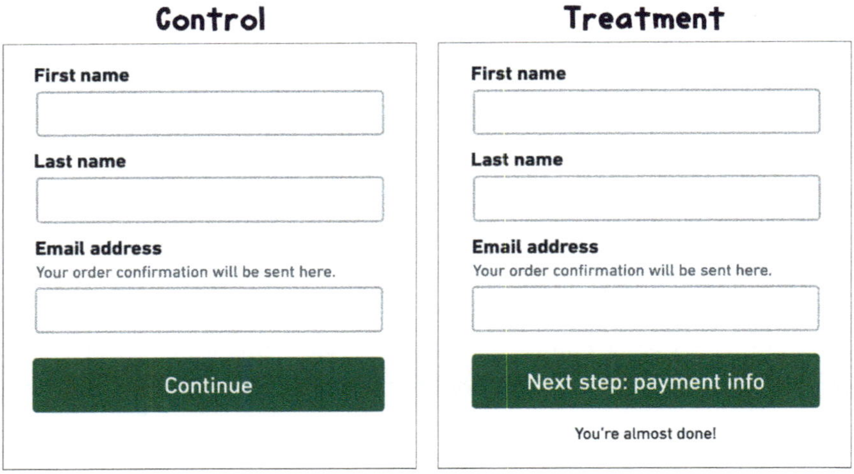

FIGURE 5.30 In the control (left), the call-to-action is a generic call-to-action, and it's unmotivating. In the treatment (right), the next step is clear and specific, and it includes a nudge to motivate users.

The Important Bits

- Remember: Although we might not often have face-to-face interactions with our customers, our job is to serve people. Approach your experiments with the care and responsibility you would if you weren't separated from them by pixels and a computer screen.
- When you design for impact, function comes before form. Focus first on access, then usability, and then aesthetics. Aesthetics are visual information, and they should reinforce messaging, and improve usability and accessibility.
- Fogg's Behavior Model and accompanying equation (B=MAP) is a tool to remind you to solve the right problem for people, to make things accessible and easy-to-use, and finally, to make the experience motivating.
- It's virtually impossible to change just one thing in an experiment. You need to understand how things get impacted around your change and work to minimize potential points of failure so that your experiment has a chance to succeed.
- To set your experiments up for success, remember Jakob's Law: "People spend most of their time on other sites." In other words, experiences with other products set the expectations for how your product works. Stick as close to existing mental models and expectations people have unless the hypothesis behind the experiment is to explicitly break them.
- Visit the Nielsen Norman Group website and blog regularly to understand how usability and accessibility expectations change and let that research inform your design decisions.
- Experiment with psychological concepts such as framing, sequencing, and motivation, and use tools such as the Hemingway App, Hofstede's Cultural Dimensions and The Culture Factor's Country Comparison Tool, and Google Trends to set your experiments up for success.
- Just because an idea worked for someone else somewhere else doesn't mean it will work for you, on your product, for your customers, and in the way that you execute on it. Silver bullets don't exist, and experiment execution can make things fail. That's why we test.

CONVERSION DESIGN PROCESS

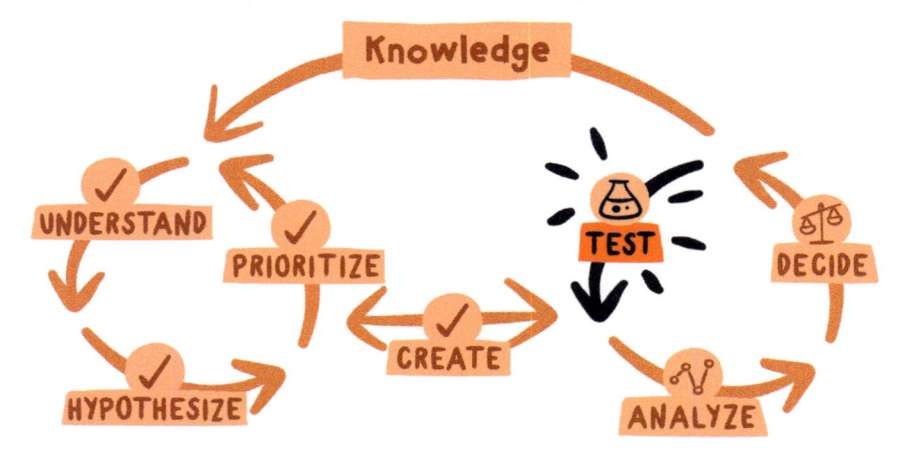

Knowledge

- UNDERSTAND
- HYPOTHESIZE
- PRIORITIZE
- CREATE
- TEST
- ANALYZE
- DECIDE

"It is not unscientific to take a guess. Although many people who are not in science believe that it is."

— Richard Feynman

CHAPTER 6

The Test Phase: Test Like You're Wrong

THE POINT

If you want to reliably measure the impact of your change, you need to design your experiment, clarify the process, and stick to it.

Two things in the world scare me: the ocean and math.[1] I've had nightmares about horrifying formulas like this $S = \sigma = \sum (x - x^-) 2n$ and this $x^- = \sum \times n$. So, I dropped math in school as soon as I was legally allowed.[2] Eventually, through osmosis from friends at work, I realized something. Behind all the scary letters, numbers, and symbols are simply words, dots, and lines.

---

1. OK—you got me. I'm also terrified of climate change and the general state of the world. I can't change that stuff. I can actively avoid both the ocean and math. But I digress.
2. I stopped taking math when I was 15. Well, that's not entirely true. I spent a year as a foreign exchange student in The Netherlands at that time. Technically, I took math that year, too. The trick is, I had no idea what was going on because it was all in Dutch. To keep a long story short, I didn't learn any math. But I did learn Dutch!

And behind each one of those things are stories and pictures. As a visual person who is OK with languages, the stories and pictures help the concepts just "make sense." So, for all y'all linguistic and visual people out there, this chapter has lots of stories and pictures to both "show and tell" what the scary stuff means. But don't worry if your brain starts to feel "melty." That's normal.

But First, a Reminder

Grab the Validate Template of the Good Experimental Design toolkit in the Appendix. It guides you through the next steps. Although you spent a lot of time perfecting your hypothesis and designing a change you believe is better, you must now shift your mindset. "Test Like You're Wrong" is the essence of null hypothesis testing. You need to assume that your change has no impact unless you get some very convincing evidence to prove otherwise. The first section of this template reminds you of that fact. It's conveniently called the "Null Hypothesis Reminder," as shown in Figure 6.1.

Null Hypothesis Reminder

The change is tested against the current state. The assumption is that the change has no effect.

However, if the effect outlined in Metrics & Math is observed, we will...

- change our minds,
- reject the current state, and
- adopt the change as the new current state.

Otherwise, we will reject the change and keep things in their current state.

FIGURE 6.1 The concept of a null hypothesis is an important part of A/B testing.

The Null Hypothesis Reminder begins: "The change is tested against the current state. The assumption is that the change will have no effect." The current state (aka *base* or *control*) is the currently accepted theory and your default belief. Your change (aka the *variant* or *treatment*) is the thing you designed to be better. Figure 6.2 reiterates that you need strong evidence to convince you that the alternate hypothesis is indeed better.

> **Null Hypothesis Reminder**
>
> The change is tested against the current state. The assumption is that the change has no effect.
>
> ==However, if the effect outlined in Metrics & Math is observed, we will...==
>
> - ==change our minds,==
> - ==reject the current state, and==
> - ==adopt the change as the new current state.==
>
> Otherwise, we will reject the change and keep things in their current state.

FIGURE 6.2 You decide in the Metrics & Math section what would make you reject the null hypothesis.

It says, "However, if the effect in Metrics & Math is observed, we will... change our minds, reject the current state, and keep the change as the new current state." Finally, there's the last line: "Otherwise, we will reject the change and keep things as they are" in Figure 6.3. This statement clarifies what you'll do if you don't observe that effect.

> **Null Hypothesis Reminder**
>
> The change is tested against the current state. The assumption is that the change has no effect.
>
> However, if the effect outlined in Metrics & Math is observed, we will...
>
> - change our minds,
> - reject the current state, and
> - adopt the change as the new current state.
>
> ==Otherwise, we will reject the change and keep things in their current state.==

FIGURE 6.3 If you don't see the effect written in the Metrics & Math section, you'll stick with the null hypothesis.

Boiled down, all the Null Hypothesis Reminder section says is, "If we see evidence the change is better, then we'll keep it. If not, we'll keep things the way they are." It might seem basic to spell out your logic like this, but it's an important part of process hygiene. Process hygiene and why it matters gets covered in the "Keep Your Math Squeaky Clean" part of this chapter.

The Nosebleed Section

The Metrics & Math bit is tricky. So, don't worry if you (possibly) get a nosebleed while reading this. Check out the Metrics & Math section in Figure 6.4.

Metrics & Math

This test is designed to find an impact on...

[_____] goal metric at a...
[____]% minimum detectable effect at a...
[____]% significance level and a...
[____]% statistical power after...
[_____] visitors, and...
[_____] run time.

FIGURE 6.4 This might look a little scary, but once you understand the concepts—and know which calculator to use—it will be fine.

That is a *lot* of blanks to fill in. These blanks represent how long you'd need to wait and what you'd need to see to be convinced that the change is better than the current state or not. The good news is that you should already have much of what you need to fill these in if you prioritized correctly.[3] In total, there are six blanks—four of these things you choose, and two of them get calculated based on your choices. The four things you choose are:

- **Goal (aka primary metric).** What gets measured to determine the impact from the change you designed.
- **Minimum detectable effect (MDE).** The smallest impact from the change you care to measure on your goal.

---

3. Which I'm absolutely sure you did. Amiright?

- **Significance level.** How many times you're willing to believe a change had an impact when, in reality, it hasn't.
- **Statistical power.** How often you want a test to find a real impact if it indeed exists.

The things that you calculate based on the things you choose are:

- **Sample size.** The number of people a change needs to be exposed to so that you can reliably measure the minimum detectable effect (MDE) on the goal.
- **Run time.** How many weeks (or months) an experiment needs to be "live" with the change exposed to visitors to reach the sample size.

If you don't understand all that after the first read—it's cool. People go to university to get PhDs in this stuff. As John Fuller, prolific poet, novelist, and children's writer once said, "All things are difficult before they are easy." So, keep going! You'll eventually get the hang of it.

> **TIP** ADJUSTING YOUR RUN TIME TO REACH YOUR SAMPLE SIZE
>
> Run-time calculations are another educated guess. It's possible that your sample size isn't achieved during your initial calculated run time. If this happens, add another full business cycle (or two) onto your run time so that you reach your sample size. The result is only valid when you hit your required sample size for significance and full cycles of data are represented. ∎

Choosing Your Goaaaaal!

Choosing your experiment goal isn't quite as exciting as watching Pelé score a goal in the FIFA World Cup quarter final—but it comes pretty close.[4] There's an empty space (the highlighted blank in Figure 6.5) in the Metrics & Math template that needs a goal, and you're the Messi[5] of this moment.

---

4. Pelé, arguably the best soccer player of all time, is the only soccer player's name I know. (OK, I also know Messi. I mention him in the next sentence. And I know David Beckham because he married Spice Girl, Victoria Beckham.)
5. Messi is a prominent soccer (aka European "football") player at the time of writing this book.

```
Metrics &      This test is designed to find an impact on...
Math
               [_____] goal metric at a...
               [_____]% minimum detectable effect at a...
               [_____]% significance level and a...
               [_____]% statistical power after...
               [_____] visitors, and...
               [_____] run time.
```

FIGURE 6.5 Your goal is at the heart of the Metrics & Math portion of your experiment's design.

To fill this blank in, revisit your Hypothesize template that you fill in during the Hypothesize phase. You can find the Hypothesize template in the Good Experimental Design appendix. Check out the Prediction section of that template in Figure 6.6. It covers the experiment variables.

```
Prediction    If we [proposed change] to
              [independent variable(s)] then
              [expected impact] on [dependent variable(s)].
```

FIGURE 6.6 Your prediction helps you choose your goal (aka primary metric).

Right now, you're probably thinking, "WHAT?" because the prediction template isn't even an actual sentence. You have to make it a complete thought. To make this concrete, imagine your customers struggling to find your product's FAQs. You think that if you turn the FAQs text link into a button that people will see it better and click on it more. You write it in the template like this: "If we change the FAQ's text link to a button, it will catch people's attention. The effect is that people will see it better and click on it more."

The independent variable is the entry point to the FAQs. It is either the text link (the current state) or the button (the change). The expected impact of the change is that people will see the entry point better. Then the dependent variable, the number of clicks on the entry point, will increase due to its better visibility. Number of clicks on the entry point is also your guiding metric (aka *leading indicator*). To choose your goal, follow that same thread of logic between the independent variable, effect of the change, and the outcome on the dependent variable shown in Figure 6.7.

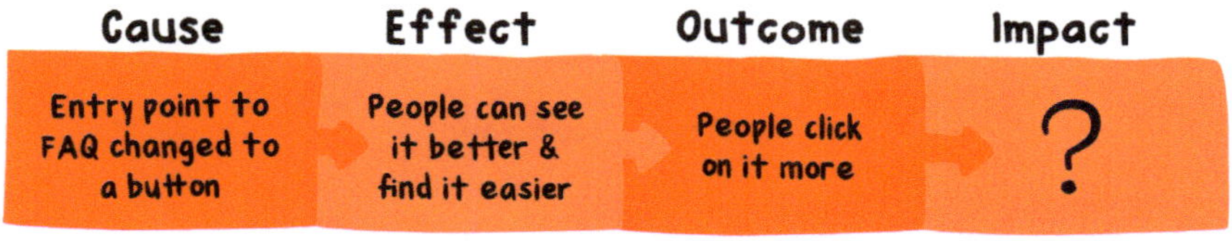

Cause	Effect	Outcome	Impact
Entry point to FAQ changed to a button	People can see it better & find it easier	People click on it more	?

FIGURE 6.7 The desired impact of your change is your goal.

Now, go a step further and ask yourself "Then what?" You need to consider the final desired impact of the change. Because customers can now see the entry point, they will find the answer to their question easier and no longer contact customer service, as shown in Figure 6.8.

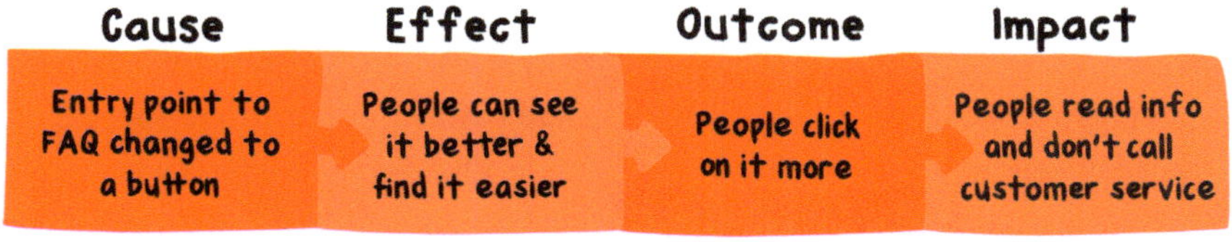

Cause	Effect	Outcome	Impact
Entry point to FAQ changed to a button	People can see it better & find it easier	People click on it more	People read info and don't call customer service

FIGURE 6.8 Clicks don't matter to users and the business, but calling customer service less does.

Thus, the goal for the Validate template is "to decrease customer service contacts." Now that you can use your prediction to follow logic to an impactful goal, document it, so what you aim to impact is clear.

Minimum Detectable Measuring Tape

Imagine that you and your friend want to know who can jump higher. So, you decide to have a jumping contest. To measure the jumps, you go to the wizard shop and buy a magic measuring tape. Although it's magic, it can only measure jumps that are at least two inches different (check out Figure 6.9).

Yes, you could have bought one that was more sensitive, but it was very expensive. So, you settled for the two-inch one. That means that if you and your friend jump and the difference is less than two inches—the tape isn't sensitive enough to report by how much. That makes two inches your game's minimum detectable effect (MDE).

FIGURE 6.9
Your magic measuring tape isn't terribly sensitive. It only logs differences of *at least* two inches.

The same is true with online experiments. An MDE is the smallest difference that a test can reliably measure. If the difference is smaller than the MDE, there's a good chance you won't be able to reliably tell what the difference is.

Figuring Out if a Test Is Worth Running

In many businesses, it takes a lot of time and resources to design and run experiments. So, you should do what you can to ensure that you're spending your time and money in the best way possible. One way to know if a test is worth prioritizing is to do some quick calculations that ultimately lead you to your MDE. You should have these numbers from the Prioritize phase. The Impact Effort matrix in Chapter 4 had you doing back-of-the-napkin calculations to determine impact. These calculations are estimations based on your best guesses informed by different types of evidence in the hierarchy of evidence. Questions to help you calculate impact and effort are:

- **How many employees will it take to run this experiment?** This helps you understand how many resources are needed.
- **How long will they need to work on it?** This nails down the number of days, weeks, or months you think it will take to design and run the experiment (aka "time spent"), because time is money.
- **How much does the required employees' time cost?** This helps you calculate the monetary investment of running the test.

- **What's the average value of the goal the experiment aims to improve?** This clarifies how much money you expect to earn or save if the hypothesis is true.
- **How much of an improvement is needed to "pay back" the cost of the labor to do the experiment?** This helps you understand the monetary break-even point of running the test.
- **How much of an improvement is enough to buy a cake to celebrate?** This helps you figure out your cake threshold.

At this point, you're probably thinking, "Wait... I need to consider cake in my calculations?" And the answer is yes. Cake is important.

Calculations Are a Piece of Cake

Cake[6] is a great way to define your minimum detectable effect. The "Cake Principle" was created and coined by Lucas Bernardi, a principal data scientist and online experiment expert. The Cake Principle is a practical (and delicious) concept to determine the smallest effect you'd care to measure. To find your cake threshold, simply ask yourself, "What's the smallest effect I'd need to see to be excited enough to buy some cake?" Then guess the likelihood that the change you propose could have such an effect. If you think your hypothesis has the potential to make that level of change, then do the experiment. If not, then maybe pick a different one. The trick is, you can never really know up front the impact that a change will have. You simply have to guess.

> **TIP GUESS LIKE A SCIENTIST**
>
> Richard Feynman, one of the greatest scientists of the 20th century, is attributed with saying: "It is not unscientific to take a guess, although many people who are not in science believe that it is."[7] It might feel unnatural to have to just "guess" in a process that's supposed to bring a bit of certainty. The good news is that it won't always feel like you're swatting around in the dark. After running some experiments, you build an intuition about what's likely to move the needle and by how much. Guessing accurately is hard at first, but eventually you will calibrate your guesses to the situation. Don't be discouraged when your first dozen guesses are way off. Not knowing is the first step of learning something new. ■

---

6. I recommend confetti or rainbow cake (for obvious reasons).
7. Richard Feynman, "The Scientific Method-Richard Feynman," YouTube, January 20, 2013, www.youtube.com/watch?v=OL6-x0modwY&t=10s

Calculating Your Sample Size and Run Time

Once you make your educated guess, then your MDE is ready to be documented in your Experiment Design documentation. (See the Validate template in the Good Experimental Design toolkit within the appendix.) You now also have a key number you need to calculate your sample size and, ultimately, run time.

> **WHY 4.6 STARS ARE BETTER THAN 5**
>
> Humans have a pretty good instinct for sample size. A common example of this is product ratings. As shown in Figure 6.10, most people would be skeptical of the higher-rated product because of the relatively lower number of ratings.
>
> But why is this? A couple of reasons are likely. First and most likely is that some amount of variation around the "real" value is expected. With 19 ratings, the "real" value for 5 could be a 4—or even a 3. But with more than 2,000 ratings, that 4.6 might be a 4.5 or a 4.7—but never a 3. The larger sample size means that we expect the number to be closer to the "real" value.
>
> The second possibility is about sample diversity. When more people rate an item, the likelier it is that the rating will be representative of the overall experience. In other words, the data is less likely to be skewed one way or another when there's a large, mixed set of responses. Whatever the reason, humans just kind of "know" that it's riskier to buy a 5-star item with 19 ratings than it is to buy a 4.6-star one with more than 2,000.

Jump Up, Jump Up, and Increase Your Sample Size

You and your friend decide to go ahead with your jumping contest from earlier. In spite of the two-inch detection limitation, you both jump four times. These four jumps are the sample size in your contest. In terms of an online experiment, sample size is how many people ("subjects" or "participants" if you're into science lingo) are in a test. The results of your jumping contest show that two out of four times your friend jumped more than two inches higher than you. "It's a fluke! We can't learn anything from four lousy jumps," you claim. "Let's measure a *lot* of jumps, so I know you're not just lucky." You've just suggested increasing your contest's sample size because you want more trustworthy data.

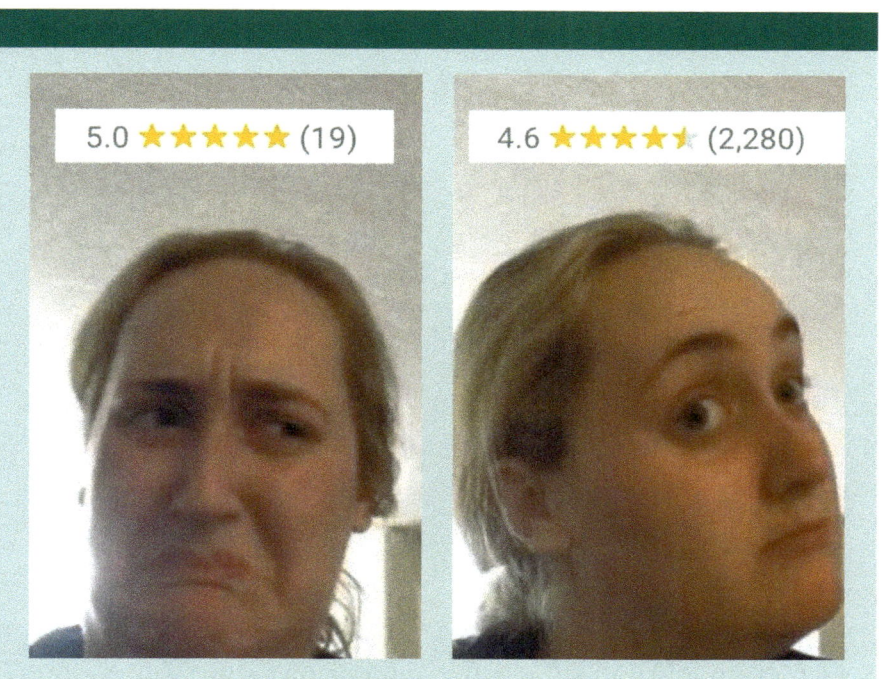

FIGURE 6.10 This meme illustrates the importance of sample size in order to consider a result trustworthy.[8]

8. Lord Red Tiger, "r/meirl," July 1, 2021, Reddit, www.reddit.com/r/meirl/comments/obkjwl/me_irl/

For 30 Days and 300 Jumps

This jumping contest has really got you annoyed. Your skepticism of the result shows that the four-jump sample size is unconvincing. So, your friend and you agree to measure 10 jumps every day for the next month. In testing speak, that makes your sample size 300 jumps, and your run time is one month (about 30 days). Run time is how long an online experiment is active and accepts new visitors.

Choosing 300 jumps and 30 days is all well and good for a jumping contest between friends. But in online experimentation, you don't just pull those numbers out of the sky. If you want a statistically sound and unbiased result, you need to calculate these numbers. As you design your experiment, the things that you need to calculate sample size and run time are shown in Figure 6.11.

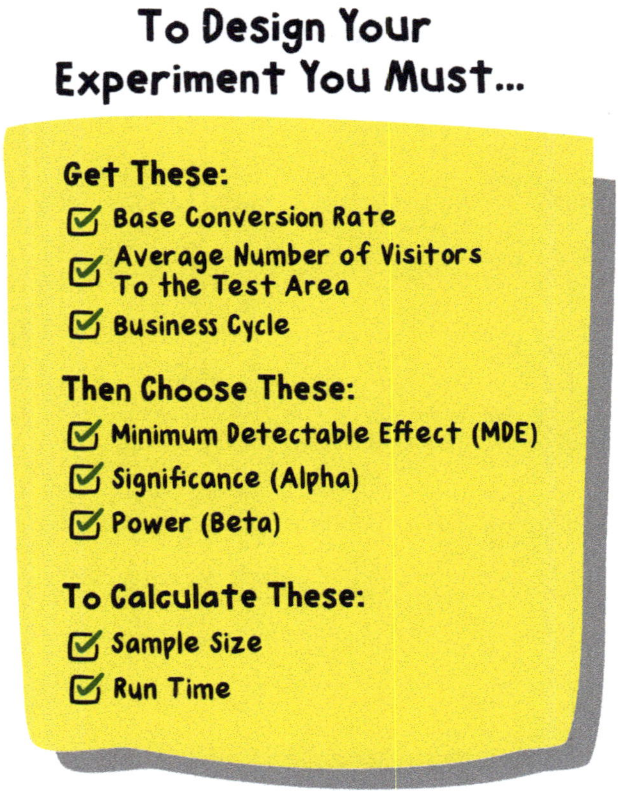

To Design Your Experiment You Must...

Get These:
- ☑ Base Conversion Rate
- ☑ Average Number of Visitors To the Test Area
- ☑ Business Cycle

Then Choose These:
- ☑ Minimum Detectable Effect (MDE)
- ☑ Significance (Alpha)
- ☑ Power (Beta)

To Calculate These:
- ☑ Sample Size
- ☑ Run Time

FIGURE 6.11 Some of the things you get from your product data while others you choose based on the situation.

Evan Miller's Awesome A/B Testing Tool

Since you knew your goal from the Hypothesize and Prioritize phases, you should have dug into your analytics or set up some A/A tests to get "the lay of the land." In this "lay of the land," you can get the conversion rate of your experiment goal as well as the number of visitors to the test area in a given timeframe. Once you have your base conversion rate, it's time to whip out a special calculator. Evan Miller's Awesome A/B Testing Tools include a free and reliable sample size calculator in Figure 6.12.[9]

Question: How many subjects are needed for an A/B test?

Baseline Conversion Rate: 20 % — 20%

Minimum Detectable Effect (MDE): 5 % — 15% – 25%

The Minimum Detectable Effect is the smallest effect that will be detected (1-β)% of the time.
● Absolute ○ Relative
Conversion rates in the gray area will not be distinguishable from the baseline.

Sample size:
1,030
per variation

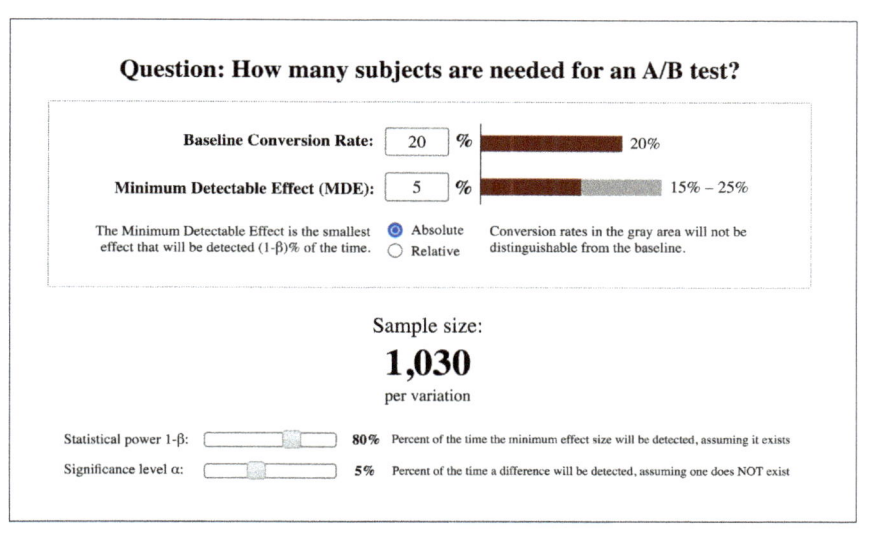

Statistical power 1-β: 80% Percent of the time the minimum effect size will be detected, assuming it exists
Significance level α: 5% Percent of the time a difference will be detected, assuming one does NOT exist

FIGURE 6.12 Evan Miller's sample size calculator is a useful tool to find your sample size. It pre-populates numbers that you need to change.

Notice that the form fields and sliders on the calculator correspond with the blanks in the Validate template. Figure 6.13 shows which blank goes with which part of the calculator's UI.

---

9. Visit www.erindoesthings.com for a link to this calculator. I have absolutely no idea how this calculator works. I assume that if I put in a good guess for an MDE and an accurate base conversion rate that magic happens, and the right sample size comes out. I asked Lucas Bernardi, my principal data scientist friend, if I can trust the math, and he said, "Yep." He uses it to double-check his own calculations. His judgment is good enough for me. Sometimes, you gotta know what you don't need to know, too!

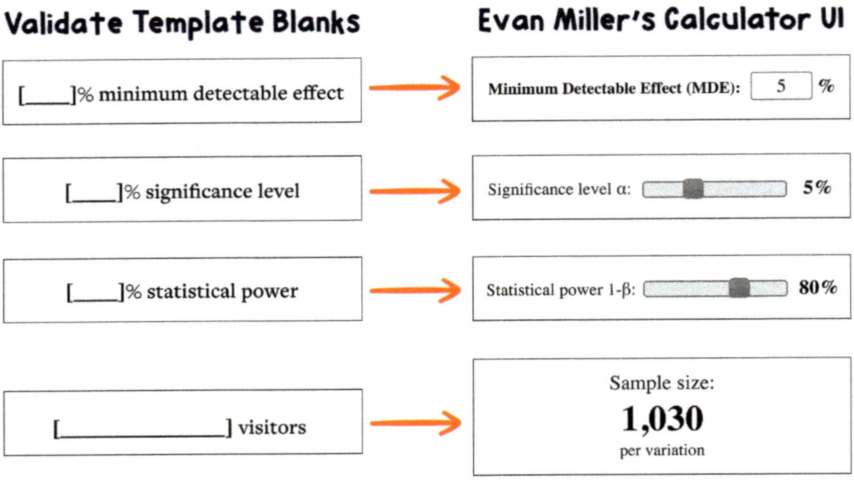

Validate Template Blanks

[____]% minimum detectable effect

[____]% significance level

[____]% statistical power

[_____] visitors

Evan Miller's Calculator UI

Minimum Detectable Effect (MDE): [5] %

Significance level α: [====] 5%

Statistical power 1-β: [========] 80%

Sample size:
1,030
per variation

FIGURE 6.13 The blanks in the Validate template correspond with the UI in the sample size calculator.

At this point, type in your base conversion rate and your minimum detectable effect. Do yourself a favor and navigate on over to Evan Miller's calculator and take it for a spin. (Ignore the sliders for now.) Play around with the MDE and conversion rate fields. As you do, notice that a small MDE needs a larger sample size. Conversely, a larger MDE needs a smaller one.

The illustration only shows what happens when you mess with the baseline conversion rate. Notice how the sample size required varies as you change it. Playing with these numbers helps you build an intuition for how sample size behaves as these inputs change (Figure 6.14).

Play With the Calculator To Learn How It Behaves

Question: How many subjects are needed for an A/B test?

Baseline Conversion Rate: 2 % — Low base conversion rate. ▮ 2%

Minimum Detectable Effect (MDE): 5 % ▮▮▮▮ 0% – 7%

The Minimum Detectable Effect is the smallest effect that will be detected (1-β)% of the time.
● Absolute
○ Relative

Conversion rates in the gray area will not be distinguishable from the baseline.

Sample size:
160 per variation ← Fewer people needed in the sample.

Question: How many subjects are needed for an A/B test?

Baseline Conversion Rate: 50 % — High base conversion rate. ▮▮▮▮▮ 50%

Minimum Detectable Effect (MDE): 5 % ▮▮▮▮ 45% – 55%

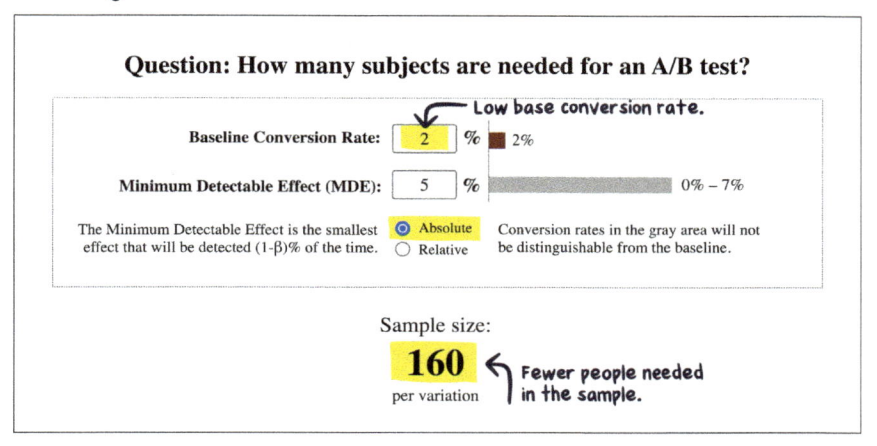

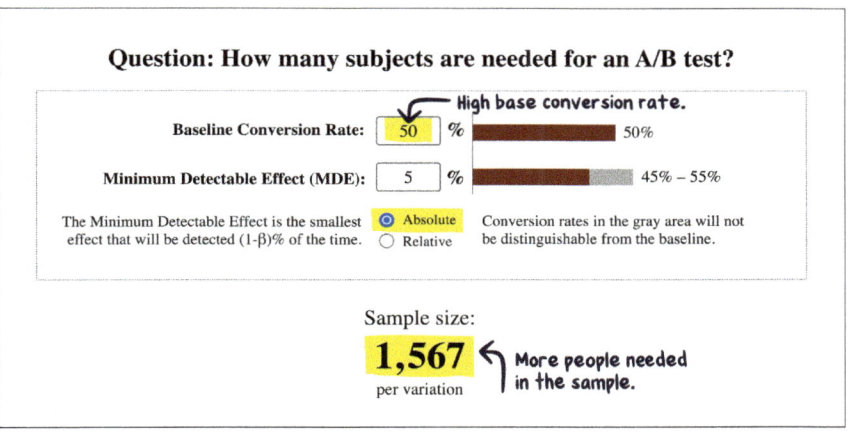

The Minimum Detectable Effect is the smallest effect that will be detected (1-β)% of the time.
● Absolute
○ Relative

Conversion rates in the gray area will not be distinguishable from the baseline.

Sample size:
1,567 per variation ← More people needed in the sample.

FIGURE 6.14 Notice how the numbers change as you mess with the inputs to understand how they relate to one another.

After you're done playing with those numbers, move on to mess around with some radio buttons.

Numbers Are Absolutely Relative

Underneath the baseline conversion rate and minimum detectable effect fields is an option to select either "absolute" or "relative," as shown in Figure 6.15.

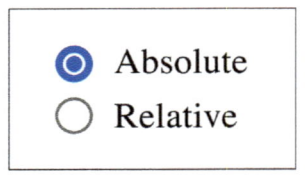

FIGURE 6.15 What this radio button is set to makes a *big* difference.

When you calculate in "absolute" terms, it means you're not comparing it to anything else. You simply use the exact amount. In this case, the exact amount is the percentage difference you want to find on your goal. For example, imagine that your base conversion rate is 10% and you want to be able to see a minimum change of 2%. That means the sample size it calculates is based on detecting an effect when the metric hits either 8% or 12%. It needs to change by two complete percentage points in either direction to hit the threshold as highlighted in Figure 6.16.

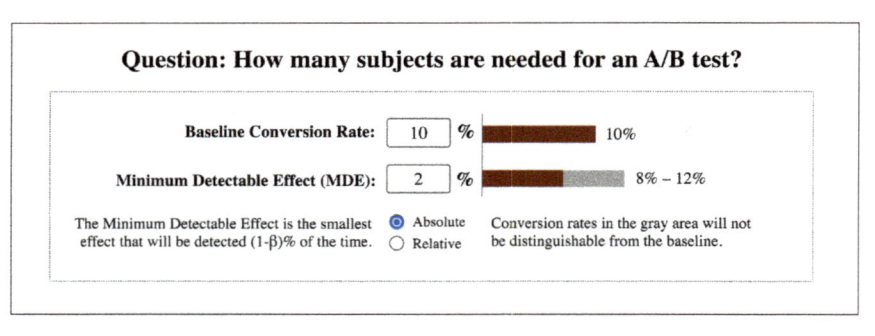

FIGURE 6.16 In the MDE row, it shows you the sample size will find an effect between 8%–12% if you choose an absolute MDE of 2% on a baseline conversion rate of 10%.

Unfortunately, you don't get any cake for making the thing worse. But maybe not making something accidentally awful is reward enough. The math to find the absolute change on your goal looks like Figure 6.17:

> **10% + 2% = 12 if the change increases the rate**
> **10% - 2% = 8 if the change decreases the rate**

FIGURE 6.17 In this case, 2% means 2% in whole number terms. The 2% is not related to anything else. It is what it is—2%.

A relative change, on the other hand, considers the current context. To find the new conversion rate after the increase, you want to calculate 2% of the original conversion rate (10%) and then add that amount to the original rate. The calculation for a relative change is shown in Figure 6.18.

> **2% of 10% = 0.02**
> **0.02 × 10 = 0.2**

FIGURE 6.18 In this case 2% means 2% in terms relative to the base conversion rate. It isn't actually two whole percentage points.

It breaks down as follows: the 2% of 10% = 0.02 part is how you get rid of the % sign to do the rest of the math. Then you take the 0.02 (which is equal to 2%) and multiply it by your base conversion rate. In this example, the base conversion rate is 10. The 0.02 × 10 = 0.2 part shows the result. This means that 0.2 is the relative increase of 10%. The absolute value of 2% (which is 2) is *much* bigger than a relative value of 2% (which is 0.2). That's why there's such a *massive* difference in how many people are needed in relative changes vs. absolute changes. The effect you look for in absolute terms is bigger than in relative terms. This is why the numbers change so much when you toggle your selection from absolute to relative, as shown in Figure 6.19.

An Absolute Change of 2%
You Actually Want To Detect 2%

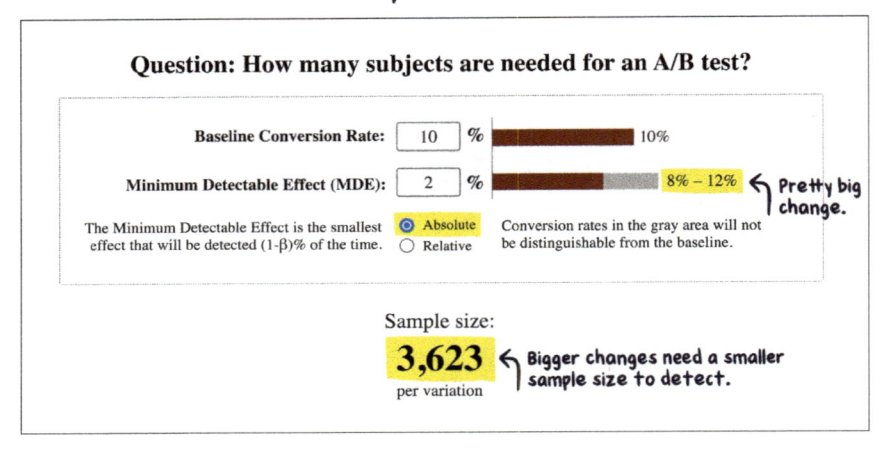

Question: How many subjects are needed for an A/B test?

Baseline Conversion Rate: 10 % — 10%

Minimum Detectable Effect (MDE): 2 % — 8% – 12% ← Pretty big change.

The Minimum Detectable Effect is the smallest effect that will be detected (1-β)% of the time.
◉ Absolute   ○ Relative

Conversion rates in the gray area will not be distinguishable from the baseline.

Sample size:
3,623 per variation ← Bigger changes need a smaller sample size to detect.

A Relative Change of 2%
You Want To Detect .2, Which Is 2% Of 10%

Question: How many subjects are needed for an A/B test?

Baseline Conversion Rate: 10 % — 10%

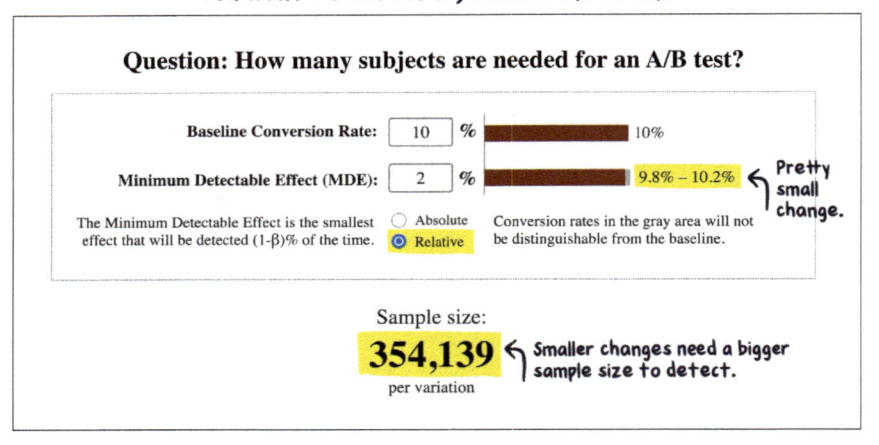

Minimum Detectable Effect (MDE): 2 % — 9.8% – 10.2% ← Pretty small change.

The Minimum Detectable Effect is the smallest effect that will be detected (1-β)% of the time.
○ Absolute   ◉ Relative

Conversion rates in the gray area will not be distinguishable from the baseline.

Sample size:
354,139 per variation ← Smaller changes need a bigger sample size to detect.

FIGURE 6.19 If you want to find an absolute change of 2% you need 3,623 visitors in your sample, which is reasonable. If you want to see a relative change of 2%, which is a much smaller effect, you need a whopping 354,139 visitors.

Now, mess with the base conversion rate. Notice that the lower the base conversion rate when "relative" is selected, the higher the sample size will be. Conversely, the higher the base conversion rate, the lower the sample size is. Playing with these numbers helps you build an intuition for how sample size behaves as these inputs change. All of this is to say: Make sure that you choose the right option. The difference between absolute and relative is a *big* deal.

Ain't Nobody Got Time for That

Low sample size can lead to untrustworthy experiment results. Some academic guidance states you need at least 100,000 visitors per variant for a believable result.[10] For many businesses, getting a number that high would take *forever*. This guidance comes from a *big* business environment where products get hundreds of millions (sometimes billions) of visitors per day. The effects they're trying to detect are small because the products at these companies have been optimized for years. This means all of the changes that could make a huge detectable effect (the "low-hanging fruit" if you will…) have already been made. The only things left are the "small" and hard-to-detect changes. In that setting, the academics working there can lean into theory because it's possible and necessary.

Most medium-to-small-sized companies have to lean into what's practical. Often, they shouldn't (and usually can't) follow that guidance because it stems from a different situation. At these companies, there's usually a lot of "low-hanging fruit" since products haven't been optimized. In these cases, there's no need for millions, hundreds of thousands—or even thousands of visitors—to make evidence-based decisions. There are lots of levers to pull to make testing possible in most products. From focusing on areas with a high conversion rate, to lowering your significance level, to making large design changes—you can shift your approach to the situation. In the end, you're the one who decides the impact you care about. And you do that with your cake threshold and your risk tolerance in your specific situation. There is a trade-off to adjusting your sample size. That is, that you increase your error rate. To understand error rates, head this way into a very confusing matrix of possibilities.

---

10. Ron Kohavi, Diane Tang, and Ya Xu, suggest this as a "rule of thumb" in their book *Trustworthy Online Controlled Experiments* (Cambridge, UK: Cambridge University Press, 2020).

The Confusion Matrix

The appropriately named *confusion matrix* (aka *error matrix*) helps people understand experiment outcomes and errors. In any experiment, two values exist:

- **The real value.** This is the impact you want to know with certainty. Unfortunately, you can never really know the "real value" of an effect, which is why you have to run an experiment to try to predict it.

- **The predicted value.** This number is what you think the effect could be, based on the data in your experiment. It's your best educated guess.

Each value line in the matrix—real and predicted—has two possibilities. Those two possibilities for each value create four combined outcomes within the matrix, as shown in Figure 6.20.

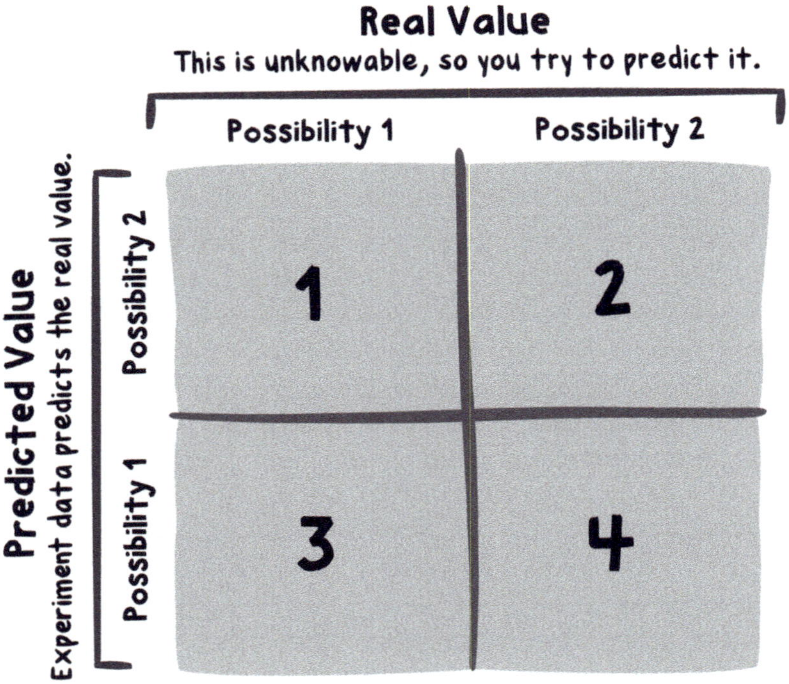

Real Value
This is unknowable, so you try to predict it.

	Possibility 1	Possibility 2
Possibility 2	1	2
Possibility 1	3	4

Predicted Value
Experiment data predicts the real value.

FIGURE 6.20 These are the two axes of the confusion matrix: real value and predicted value.

The real value is represented by the horizontal line. It can be one of two options: "true" or "false." True and false deal with a concrete reality. Reality is expressed in absolute terms because the question has a clear (albeit unknowable) answer. The true/false axis concretely answers the question, "Is there a real effect?" If the answer is "Yes, the change had a real effect," the experiment goes in the "true" column. If the answer is "No, the change had no real effect," it goes in the "false" column.

The predicted value is represented by the vertical line. Predicted values also have their own fancy way of saying "yes" and "no." These can be "positive" or "negative." Why positive and negative? Because the answer is relative to the experiment data, not an absolute truth. These two rows in Figure 6.21 answer the question "Do you see a statistically significant effect in the data?" If the answer is "Yes, you do see a statistically significant effect," then the answer goes in the positive column. If the answer is "No, you don't see a statistically significant effect," it goes in the negative one. Figure 6.21 illustrates the real and predicted values and their associated questions.

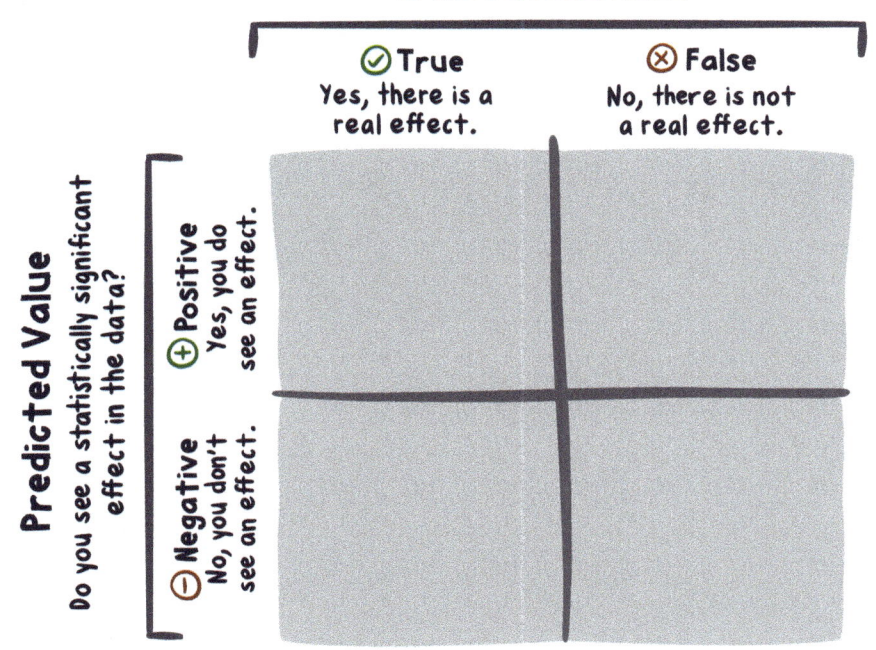

FIGURE 6.21 It's important to distinguish between the real value, which is unknowable and the predicted value, which you try to guess based on data collected from your experiment.

Yes/Yes and No/No Make a "Right"

The confusion matrix combines the different outcomes to compare reality to your prediction. You want your prediction to match reality. The ideal result is two "yeses." In two of the combinations your prediction is right, as shown in Figure 6.22.

Real Value
Is there a real effect?

⊘ **True** — Yes, there is a real effect.
⊗ **False** — No, there is not a real effect.

Predicted Value
Do you see a statistically significant effect in the data?

⊕ **Positive** — Yes, you do see an effect.
⊖ **Negative** — No, you don't see an effect.

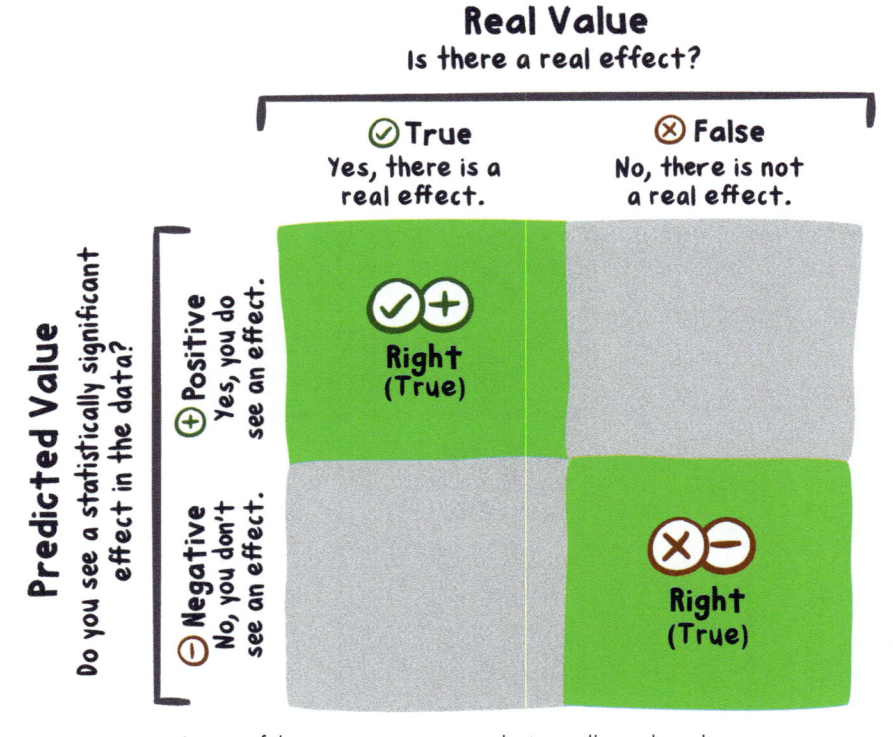

- Positive / True: **Right (True)**
- Negative / False: **Right (True)**

FIGURE 6.22 In two of the outcomes, your prediction will match reality.

The first right outcome is called a *true positive*. It's when the prediction you made matches the real value. You found an effect, and it's real. The second right outcome is called a *true negative*. This is when you observe no effect in your data, and there is indeed no real effect. If this confuses you because the "x" and the "-" look negative, that's because it's a *double-negative*. Double negatives happen when two negative things cancel each other out and make a positive. When your prediction matches reality—regardless of if it's a "yes" or a "no"—it is still "true." Both types of true errors are illustrated in Figure 6.23.

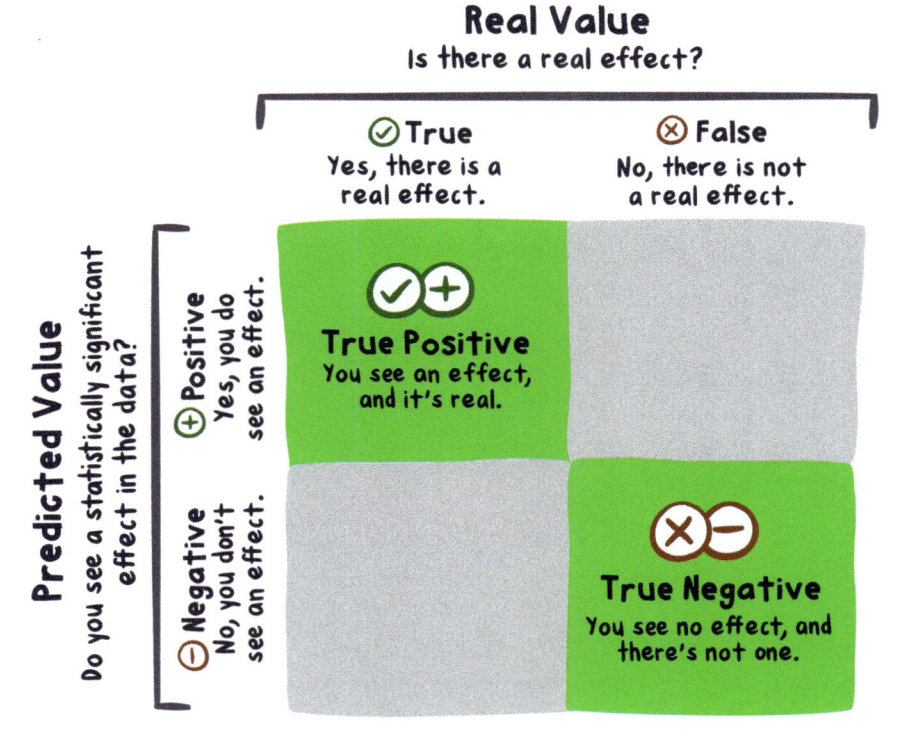

FIGURE 6.23 Both a true positive and a true negative are when the real and predicted values align.

Yes/No and No/Yes Make a "Wrong"

In the two wrong outcomes, as shown in Figure 6.24, your prediction doesn't match reality.

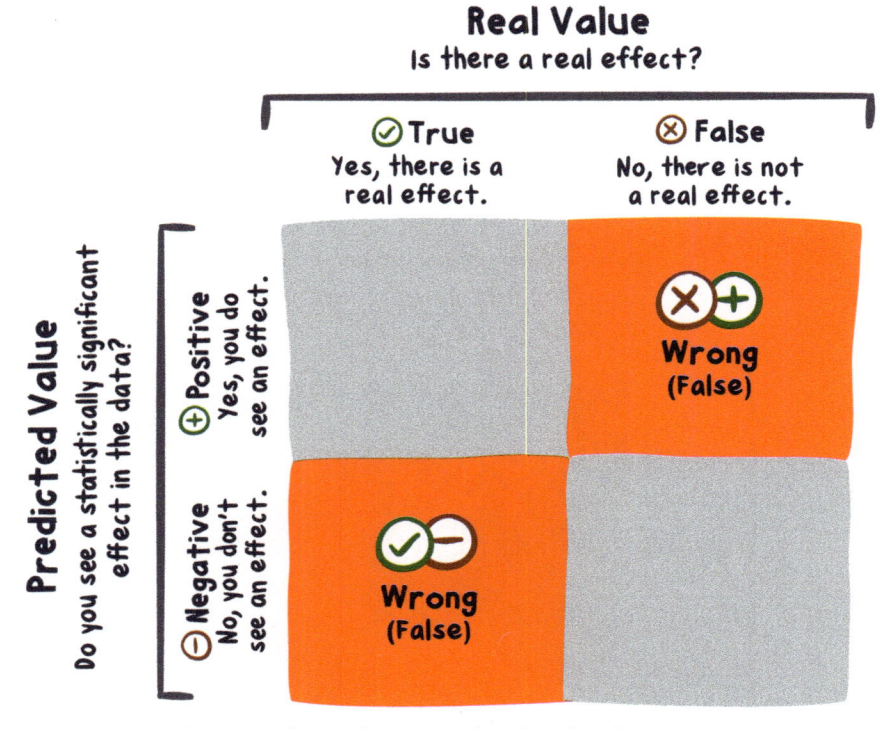

FIGURE 6.24 If your prediction doesn't match reality, then there's an error.

You can tell because it's a yes/no and no/yes combination. When this happens, it's called an *error*. You want to be aware of potential errors because they can lead to bad decisions. You decide what error rate you can tolerate with the fancy sample size calculator by adjusting the significance and power sliders. (More on how and why to do that in the upcoming section, "To Err Is Human—to Mitigate Is Divine.")

The first kind of wrong outcome is a "false positive." It's also known as a *Type I error*. This happens when your predicted value doesn't match the real value because it predicted an effect, but there isn't one. The second kind of wrong outcome is a "false negative." It's also known as a *Type II error*. This is when there is indeed a real effect. However, you don't see the effect in your data. So, you think there's no effect, when really—there is one. Both types of errors are shown in Figure 6.25.

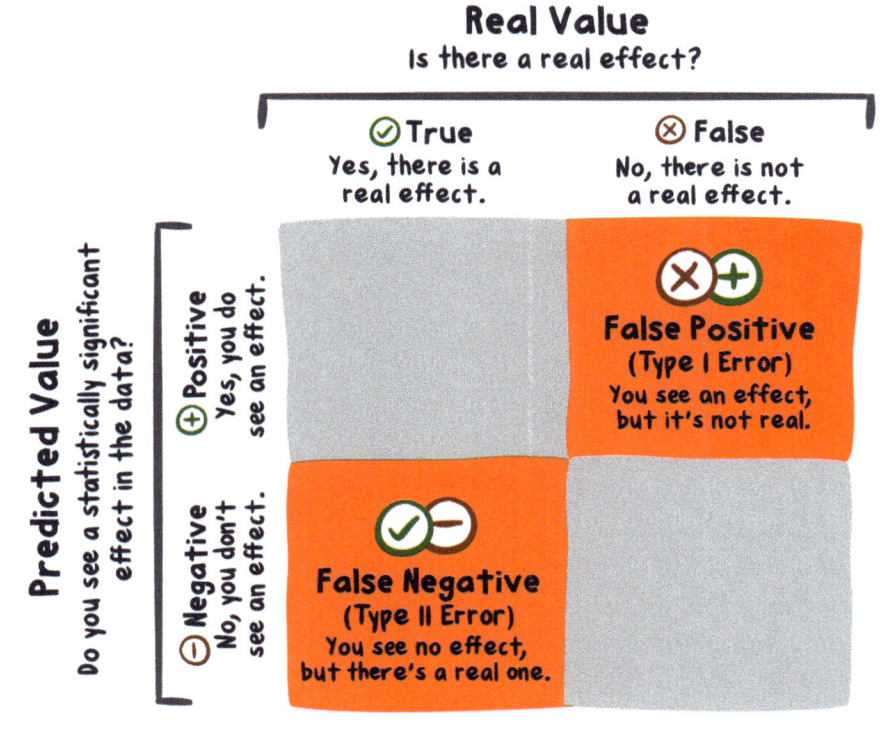

FIGURE 6.25 When the *real* and *predicted values* don't align, the results are considered *false*.

THE TEST PHASE: TEST LIKE YOU'RE WRONG · 155

Positives and Negatives That Can Be Positive or Negative

All the outcomes with either a real or predicted effect can have a positive or negative direction. This is where people typically get confused because positive and negative refer to *two different things at the same time.* The first positive or negative refers to whether or not a test *detects* an effect. The second positive or negative refers to the *direction* of the effect, which can be either greater than or less than zero.

For example, imagine that you have an experiment result that has a true positive result. The direction of the effect can be either positive or negative (see Figure 6.26).

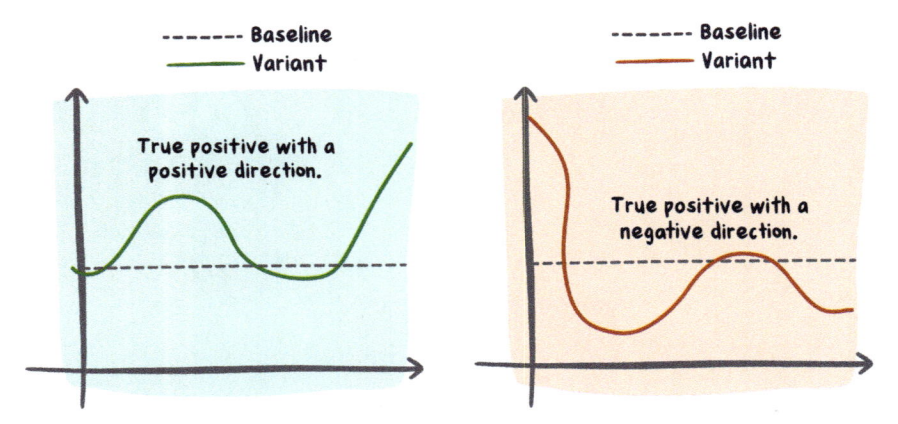

FIGURE 6.26 It's possible to observe a true positive result that has an impact that is in a negative or positive direction. This might seem counterintuitive.

When you thought about the term *true positive*, you possibly imagined an experiment with a positive result. However, it's possible to have a true positive that has a negative direction of the effect. For example, most people imagine a true positive has a positive impact on sales when they hear true positive. It's just as possible, though, that the true positive has a negative impact on sales, as shown in Figure 6.27.

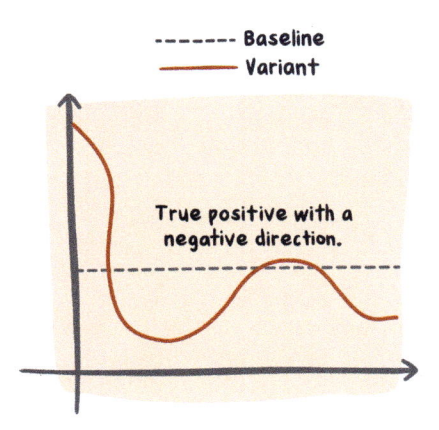

FIGURE 6.27 A true positive result that shows an effect with a negative direction.

Similar to a true positive, a false positive can have a positive or a negative effect. The example from a true positive applies here, too. Imagine that you have a false positive outcome. The effect could have a positive effect on sales or a negative effect on sales, as shown in Figure 6.28.

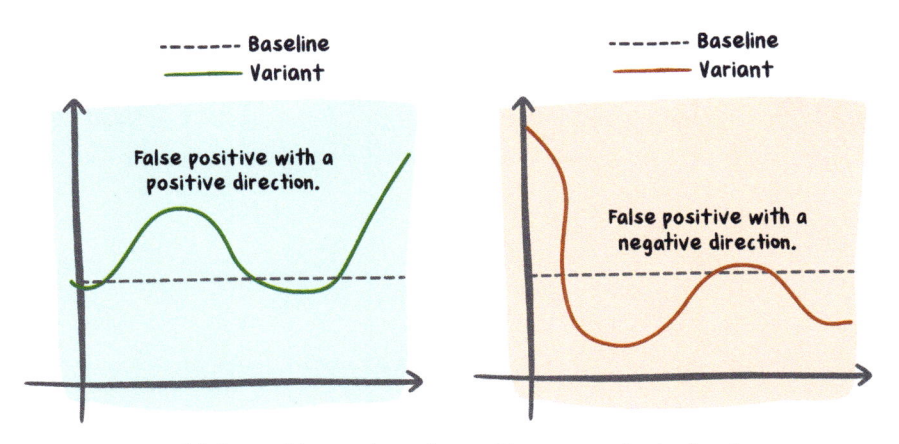

FIGURE 6.28 A false positive result can be positive or negative in direction.

THE TEST PHASE: TEST LIKE YOU'RE WRONG • 157

A true negative simply means that you will not observe or detect any effect at all. It's the only one where the direction of an effect doesn't apply because there's no effect to have a direction. Figure 6.29 sums up the four possible test results and what they mean.

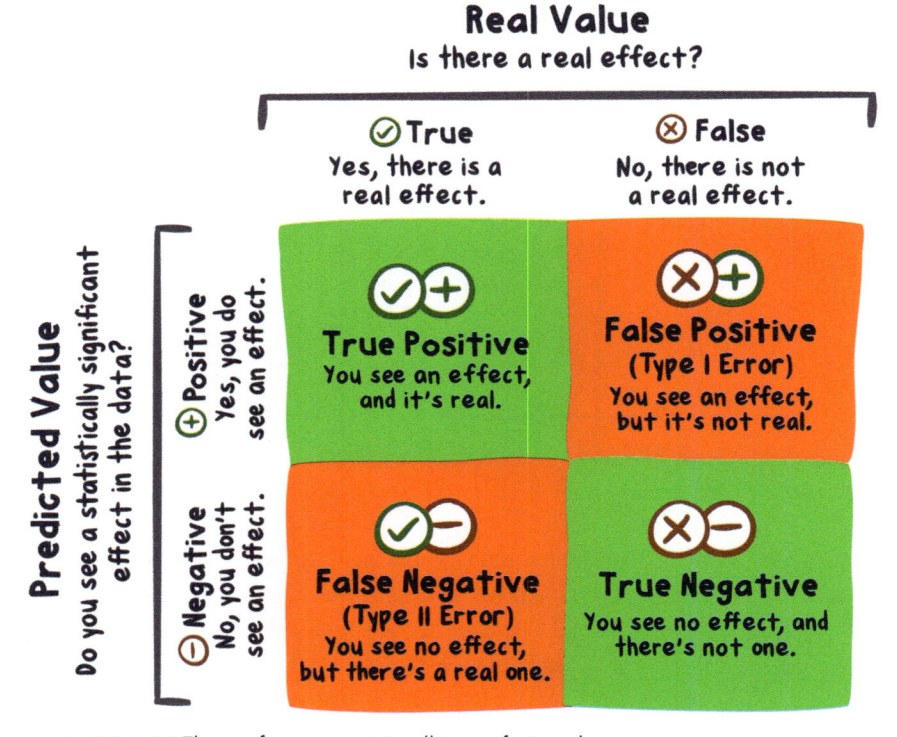

Real Value
Is there a real effect?

	✓ **True** — Yes, there is a real effect.	✗ **False** — No, there is not a real effect.
⊕ Positive — Yes, you do see an effect.	**True Positive** — You see an effect, and it's real.	**False Positive (Type I Error)** — You see an effect, but it's not real.
⊖ Negative — No, you don't see an effect.	**False Negative (Type II Error)** — You see no effect, but there's a real one.	**True Negative** — You see no effect, and there's not one.

Predicted Value
Do you see a statistically significant effect in the data?

FIGURE 6.29 The confusion matrix in all its confusing glory.

Do you understand why it's called the confusion matrix now? Take a few minutes to recover before you dive into the next topic: How to control an experiment's error rate.

To Err Is Human—to Mitigate Is Divine

Everyone hates to make mistakes. And a benefit of experimentation is to methodically limit errors to harness the compound effect. Statistical power and significance level directly impact your error rate. As previously stated, these two things are also needed to calculate your sample size and run time (Figure 6.30).

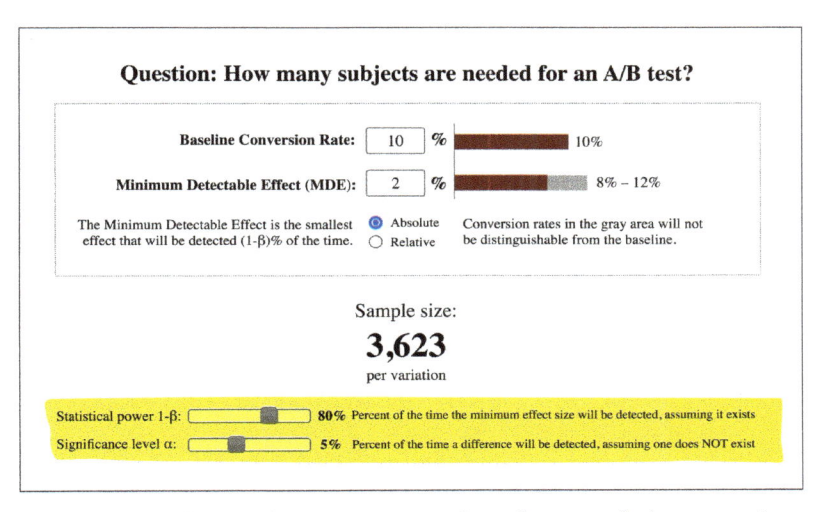

FIGURE 6.30 You'll now adjust your power and significance to find a practical sample size.

Now you can look at the two highlighted sliders you were supposed to forget about earlier. The confusion matrix might seem like a weird detour amid calculating your sample size. These sliders are how you choose your significance and power. Your goal is to strike a balance between the sample size needed to detect your MDE and your error rate. Your error rate depends on how willing you are to make a mistake, depending on how risky the experiment is.

You can follow some "rules of thumb"[11] if you don't want to think too hard about these sliders. The calculator comes prepopulated with a default value for many inputs. Many people stick with these defaults. But if you know what they do and understand the trade-off by moving them—it's super handy to adjust them according to your specific situation.

---

11. Scientists don't like the term *best practice* because it's not precise enough. They use *rule of thumb*, which means a broadly accurate guide or principle that's based on practice instead of pure theory.

I've Got the Power (and Significance)

False positives (type I errors) are closely related to your significance level. These happen when you incorrectly reject a true null hypothesis. The significance level is the threshold you set at which you decide to reject the null hypothesis. It represents the chance of observing a false positive, assuming that there really is no effect. It's also the number that's directly related to whether or not the reported result is statistically significant or not. *Statistical significance* means that the result you observe is unlikely to happen by luck because it beats the "most-likely-not-due-to-chance cutoff" you set when you chose your alpha level. The calculator's default setting for significance is 5%, as shown in Figure 6.31.

FIGURE 6.31 People refer to this value interchangeably as .05 or 5%. It's a matter of preference because they mean the same thing.

The complement to the significance level is the confidence level. This is typically 95%. This means that if you would run the same A/B test experiment over and over again, 95 times out of 100, your results would show the true value somewhere within the reported range. The other five times, the result might fall outside the reported range due to random chance. This allows you to be reasonably confident that the observed results are not just by chance. Now notice that scary symbol next to the significance level highlighted in Figure 6.32.

FIGURE 6.32 Greek letters represent concepts in math formulas. Alpha (α) represents significance.

Statisticians love math formulas. And they use symbols for ideas to keep things manageable.[12] False negatives (type II errors) relate to your statistical power. The default power is typically set to 80%, as shown in Figure 6.33.

---

12. To help my brain not melt down when I see math formulas, I now just remind myself that it's just the way a mathematician writes a sentence. Each letter symbolizes an idea, and ideas don't scare me.

FIGURE 6.33 People refer to this value interchangeably as .8 or 80%. Again, it's a matter of preference.

This means that the experiment will accurately detect your minimum detectable effect 80% of the time if it indeed exists. The other 20% of the time, it could miss the true effect. Notice the scary formula next to the words "statistical power" in Figure 6.34.

FIGURE 6.34 1-β is the formula for statistical power.

It says, "1-β" which is the formula for statistical power. Why would you want to mess with the default values then? A few good reasons are:

- Time
- Traffic
- Conversion rate
- Level of risk

Suppose that a change isn't too risky, and it doesn't require much investment. In that situation, it makes sense to tolerate more errors. Setting a slightly lower false positive rate for a copy test that gets run once and is done isn't as bad as getting a false positive on a new feature that needs lots of ongoing investment. On the other hand, suppose that a change is a maintenance-heavy new feature that requires big investment. If it's easy, low risk, and low cost—you can take more risks. However, if it's a big change, is a risk, and involves maintenance and investment, you want to be very sure it's doing what you think it is.

Another example is time. If you work at a startup with limited funding, waiting months for a result is a deal-breaker. The business situation might require more risk to get things done before money dries up. On the other hand, a company with lots of funds has time to ensure that an important guardrail metric, such as loyalty, isn't impacted by a change.

Run Time

Once you've calculated your sample size, you can figure out your run time.

It's tempting to think that the only thing to consider when calculating your run time is how many visitors per day your product gets in the area the test will run. But that's just one part of the equation. The other part is your business cycle. A business cycle is a somewhat regular pattern of user behavior. Business cycles are often measured in weeks or months to account for "seasonality." Seasonality is the idea that the time of the year, time of the day, as well as time of the week influence people's behaviors. For example, an e-commerce website might get more traffic during the holidays, and a news website might have more visitors in the morning. Capturing too many or too few of these times or days in your sample could bias your data. This can affect the outcomes of tests run during these periods. If you don't capture data in a full business cycle, it's more likely the result will be biased (i.e., kind of useless).

Run times range anywhere from one week to a few months. Shorter than this, and the results aren't reliable. Longer than this, and it's too slow to be of practical use. In fact, a few months is really pushing it. And even if you have a lot of users and you can get statistically significant results after a few days—you should still consider your company's business cycle to ensure that the data are representative.

Don't Make Things Worse

Now it's time to fill in your test type. You can run a superiority test, or a non-inferiority test. A superiority test aims to learn if there's any measurable effect between two options. It answers the question, "Is the change better on your goal than the original by more than zero?" With it, you can understand if any impact is by chance, or if there's a cause/effect relationship.

For example, imagine that you're testing a new feature. It would need the support of two new engineering teams and more customer support workers. For many businesses, that's a big investment. Business is about investing wisely, so the impact of the new feature should, at the very least, pay for itself in terms of customer loyalty, reduced marketing spent, or money made. The moment when an effect is large enough that it is net neutral in terms of money savings or sales is called the *break-even point*.

For low-risk and low-investment changes, such as usability and accessibility tweaks, it makes sense to keep a change that's better than zero. That's because there's often no additional investment required after the once-and-done improvement is made. Superiority tests help you make things better by giving you reliable information to make optimal decisions. You just have to use your judgment to decide what is "optimal" given the type of change and the situation.

A non-inferiority test uncovers wildly negative outcomes. Like a super superiority test, you decide what impact on your goal metric less than zero you think is bad enough to call *inferior.* These tests safeguard against making very bad decisions. This goal is a low bar: "Do no obvious harm." It doesn't mean that whatever gets released in a non-inferiority test has no effect. It just means that any impact there is, isn't big enough to reliably see. In reality, it could be better, or worse—you just can't know which. To run a non-inferiority test, you must decide how much of a (potential) loss you're willing to take to keep the change.

Your experiment goal determines your test type. If you want to build success with the compound effect, then run a superiority test. If you want to make a strategic leap without causing too much harm, then run a non-inferiority test. In most cases, you should run superiority tests. As a conversion designer, your goal is to make things better—not just different (or accidentally worse). Table 6.1 sums up the different test types for future reference, and the goal of your test determines the type of test you'll run.

TABLE 6.1 TYPES OF A/B TESTS

SUPERIORITY GOAL: "BETTER THAN ZERO"	NON-INFERIORITY GOAL: "DO NO OBVIOUS HARM"
To learn if one thing is better than the other one by more than zero. Helpful to create exponential growth through the compound effect consistently over time.	To learn if a change is noticeably worse than the original. Helpful to make a strategic decision to (potentially) leap over a local maximum.

WHERE "DEFAULTS" CAME FROM

One day, my friend Lukas[13] explained significance and power to me. I wanted to understand the logic behind .05 as the default of significance and 80 for power. I asked him, "Yeah, but why point zero five? And why 80? Where'd those numbers come from?"

He looked at me and burst out laughing. "Oh, you're gonna *love* this…" he said. :::dramatic pause::: "Because four times five is twenty." Of course, I had *no idea* what he was talking about. I worked hard to make a connection between what he said and what I asked, as illustrated in Figure 6.35.[14]

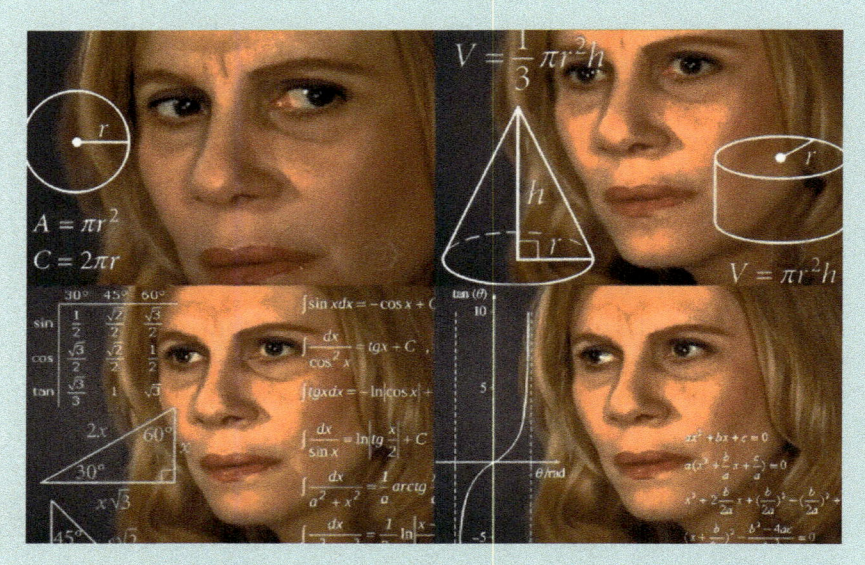

FIGURE 6.35 I had no idea what four, five, and twenty had to do with .05 and 80.

---

13. Not "electric toothbrush Lucas." Or "cake Lucas"—another one. This is Lukas Vermeer, a Dutch online experimentation expert.
14. Artur Ferreira, "Math Lady/Confused Lady," Know Your Meme, October 10, 2016, https://knowyourmeme.com/memes/math-lady-confused-lady

After a second of thought, I melted into an inconsolable puddle. I took a moment to compose myself. Then I muttered: "What?" (This was my best attempt at asking a clarifying follow-up question.) He launched into a story about some stats guy (he thought maybe it was "Fisher") who had an opinion on those numbers in the 1950s. After he finished, he said, "But don't quote me on that.[15] You probably want to look it up."

So, I looked it up. I indeed found that a dude named Fisher "preferred .05" for significance. This preference came from his belief that publishing a paper that's actually a false positive is about four times worse than missing a true discovery. (That's where the $4 \times 5 = 20$ comes from. Convert .05 to a percentage and you get 5. Multiply that by 4 because Fisher thinks a false positive is 4x worse and you get 20. Then subtract that 20 from 100 to get his preferred power of 80.) Since Fisher *preferred* it, everyone else just went along with it. Fisher was the person who created the concept of randomized controlled experimentation, so most everyone in the UK revered his opinion and didn't question his preference.

Eventually, another guy by the name of Cohen came along. He was from North America, and he wasn't starstruck by Fisher and his ideas. He believed that the alpha shouldn't be fixed (not stuck at .05 forever because Fisher "preferred" it). Instead, he said it should be a variable. A variable is something that can change depending on context. That way, you could adjust your sample size if you had a fixed number of participants to work with.

In the end, it has to do with "how much worse" making one type of error is than the other. And that depends on your context. So, the less you want to make a Type II error, that means the more you'll make a Type I error. And vice versa.

Anyway, the whole point of this is simply to say that *you decide what you care about*. Making judgment calls is a core part of science. But the magic happens when you know the risks and benefits of your judgment calls and weigh them carefully.

---

15. Hopefully, this doesn't count as "quoting" him on that.

Keep Your Math Squeaky Clean

Everything in the template is for process hygiene. Writing things down and sticking to them keeps your math squeaky clean. When you follow a pre-determined process, it guarantees the test's error rate. This provides you with trustworthy data to make optimal decisions. If you change the process partway through or after the test, the math gets thrown off. Then your error rate goes up. The trick is that you can't really know how much the error rate increases. To keep the math on your side, make your plan, document it in your Good Experimental Design toolkit, and stick to it.

The Important Bits

- To design your experiment, you must choose your goal, MDE, significance, and power up front. You use those things to calculate your sample size and run time.
- In a classic fixed-horizon A/B test, stick to one clean decision point that tells you when your sample size is achieved and your run time ends.
- If you have relatively low traffic (below 100k sample size), you can still experiment. You can adjust your significance and power to be in line with your risk tolerance, the impact you aim to have, and how much time you're willing to wait to get the sample size you need to have enough reliable data.
- Good experiment design and process hygiene control your test's error rate. The confusion matrix illustrates the different types of possible errors.
- Two common test types are superiority (make things better) and non-inferiority (hopefully not make things worse). You must decide which test type you're running before you start your experiment.
- Use the Good Experimental Design toolkit to ensure that your experiments are well thought-out, proper hygiene process is followed, and that your decisions are thoroughly documented.

CONVERSION DESIGN PROCESS

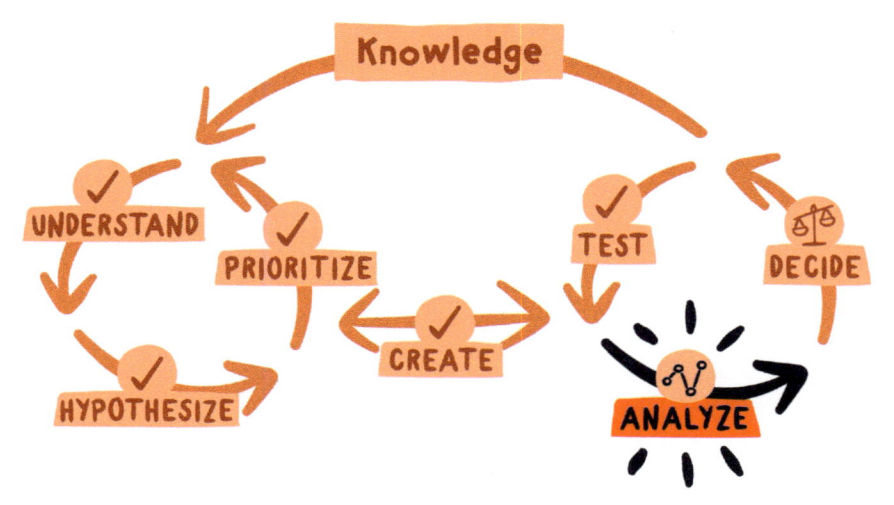

"The most important questions of life are indeed, for the most part, really only problems of probability."

– Pierre-Simon Laplace

CHAPTER 7

The Analyze Phase: Learn from the Data

THE POINT

To understand the impact of a change, you must interpret the data according to your experiment design and take care to avoid bias.

Every morning, I would sip a cup of coffee and look at my experiments.[1] One morning, I noticed a "non-inferiority bug fix" performing horribly. I made a simple change in the order of the "Favorite" and the "Add to cart" buttons as shown in Figure 7.1.

---

1. No milk. I don't want to worry about heat changing the taste like Dr. Muriel Bristol did back in Chapter 3.

FIGURE 7.1 In the control, the main call-to-action button was below the secondary call-to-action button. In the variant, the main call-to-action took the top spot.

My designer logic was that main calls-to-action should be presented before secondary calls-to-action. When I looked at the data, it was a straight red line down. I assumed this "bug fix" would have no significant impact. But I'd never seen an experiment look that bad before. It had been running for less than 24 hours, so I shrugged it off and chalked it up to randomness. Surely it would even out in a few hours. But it got worse. So, I tested it on a few different devices and browsers to see if there was a bug in my "bug fix." Everything seemed fine. At this point, I cried out to my coworkers for help: "Does anyone know why this bug fix looks so bad?"

A few minutes later, my boss, Stuart, stopped by my desk. "That's not a bug," he said. "Andrew (another designer) put that button there in a very positive experiment. Moving it changes something that matters."

I ran to Andrew's desk where he confirmed Stuart's suspicion.

"Oh yeah," he confirmed. "Don't move that button—not even a little. I tested it in a few different spots. Every time it's moved, conversion tanks."

I looked through Andrew's experiment data to learn the hypothesis behind what I considered "illogical" placement. What I thought was a bug was a useful feature that I could never imagine. When I realized that the effect was real, and more importantly—that it wasn't a bug—I turned the experiment off. "Well, that decision just passed another round of validation!" I exclaimed.

I now realize that good UX and the concept of "UX bugs" are relative to who experiences them. After many surprising experiments like this, I've been humbled. The point is: Never assume a "quick bug fix" is a "no-brainer." People use things in unimaginable ways.[2] Experiences like this taught me that design is an act of humble discovery in service of others. And analyzing data helps to uncover those impactful moments of learning.

Bad Behaviors and Data Quality

Luckily, in the opening story, I tested my "bug fix." I had lots of reliable evidence to reverse course and make a better decision than had I not tested it. That said, you can collect the most reliable data in the world, but if your A/B test analysis behaviors are rotten, your results will stink anyway. Bad data hygiene lowers data quality and ultimately decision-making quality. Many newbie experimenters (my younger self included) have done the following data analysis blunders at some point in time. The following section covers bad data analysis behaviors, why they're bad, and how to avoid them.

You Can Look, but Don't Peek

After you start your test, you can *look* at the data as much as you want. But don't *peek*. *Peeking* is when you look at the data with the intent to make a decision and you do something. *Doing something* includes things like changing the

---

2. And engineers can also introduce new bugs within bug fixes in ways we could never imagine.

sample ratio, letting the experiment run longer, or stopping it early.[3] Peeking is about intent and action. When you peek, it increases your false positive rate. As mentioned in Chapter 6 in the section "The Confusion Matrix," false positives happen when your experiment shows a statistically significant effect (either positive or negative), when one, in fact, doesn't exist. Then, from observing this data, you wrongly reject the null hypothesis so that the treatment becomes the new default. A lot of the time, peeking is not worth the risk of making a decision based on bad data. After all—the whole point of A/B testing is to de-risk decision-making. Peeking adds risk back in.

The remedy to the peeking problem is to *look* instead. *Looking* is when you make sure that everything works as expected. You have no intent to make a decision—or do anything else—with what you see. The difference between peeking and looking is intent and action. Four factors to check while looking, not peeking, are the following:

- **Check that the data is rolling in.** It's possible that something went wrong with the tracking and visitors aren't being recorded. There's nothing sadder than assuming that everything is going to plan—for weeks—only to find no data at the end. (Womp womp.)

- **Make sure that the visitors are split correctly.** It's possible that data is rolling in, but that it's not distributed according to your experiment plan. If you check for sample ratio mismatches (SRMs) manually, be careful! SRM checks are considered peeking because statistics apply to sample ratio data, too. Each time you check for an SRM, you could observe a false positive where it looks like there's an SRM, when, in fact, there isn't one. The more you check, the more you might see one. To counteract this, pick a standard moment for each test when there's enough data for a trustworthy result, but not so much data that you lose a lot of time if you do find one. Then check for an SRM once.[4]

---

3. If you're thinking, "Wait—Erin just contradicted herself! She told a story about stopping an experiment early and now says, 'Don't stop experiments early.'" You're right, I did contradict myself. *Kinda.* The difference between the story and "peeking" is that in my story I *learned something new* that explained the result. I didn't stop the experiment based purely on what I saw. The "learning something new" from the result is what made it OK for me to stop it.
4. You can also set up SRM checks with a sequential approach instead of having a single fixed-horizon. That means you can check for an SRM more than once without invalidating the math. If you're a stats nerd, check out the book *Statistical Methods in Online A/B Testing* by Georgi Georgiev, or read www.analytics-toolkit.com/glossary/sequential-testing/ for more context.

- **Find potential bugs.** Data before your preselected decision point is pretty unreliable. But you can look at it skeptically and poke around where things look interesting. Looking at the data is a great way to narrow down what conditions (for example device types, operating systems, languages, browsers, etc.) might need extra testing. If the data says that one of these things is lagging, make sure that there's nothing weird by testing your experiment in those conditions. But again, beware—you have a higher chance of observing a false positive in this data, too. Looking can occasionally help you catch very, very bad bugs that have a very strong effect. Catching and fixing edge-case bugs early is a great way to maximize your experiment's chance of success as well as save time and money.
- **Have fun and learn!** Looking at your experiments daily over your drink of choice is a worthwhile ritual because it teaches you how data behaves so you build an intuition over time. You can and should investigate when things appear bad because there might be something to learn from the observation. But that all said, don't willy-nilly stop things every time you suspect a problem without more evidence of one.

As you look at experiments, you'll quickly learn that data behaves in all kinds of weird ways.

The Choppy Seas of Data Volatility

One reason not to make snap judgments based on incomplete data is because data can be volatile. In other words—it can be all over the place. How close together or "all over the place" data is, is called *variance*.

When you've run and monitored lots of experiments, you'll notice that data often looks very positive or very negative shortly after you start a test. This volatility primarily happens due to low sample size. When there's not much data in a sample, it takes fewer data points to skew a result.[5] Each point has an exaggerated effect because there isn't much to counterbalance it.

Knowing when data is just volatile vs. actually telling you something is an intuition you build over time. This intuition comes after running many,

---

5. This is what Fisher learned in the tea experiment back in Chapter 3. There was a big difference between Dr. Bristol guessing *all* cups of tea right vs. guessing *most* cups of tea right because of the low sample size.

many, *many* tests. Eventually, you'll learn to know if what you see is likely a bug, a huge effect, volatility, or an SRM.[6] This intuition is like earning your "sea legs" for data volatility. To fix the problem of poorly navigating volatility at its root, fix your horizon.

Set Your Eyes upon the Horizon

To control your false positive rate, pick one point in the future to look at the data and make your decision. This should be done before you start your test. (I repeat: **before** you start your test.) Then stick to that decision moment. When you pick your decision moment up front, it's called *fixing your horizon* in a *fixed horizon test*.[7] The "horizon" is the timespan the test needs to run. It should be enough to achieve the sample size and account for your typical business cycle. The "fixed" refers to the point at the end of your experiment runtime when you make your decision. Write the experiment end date and time in your Validate template in the Appendix under the "Decision Moment."[8] This documented decision moment reminds you to stay the course when things get choppy.

Experiment Analysis 101

Data literacy is key to democratic decision-making in experimentation cultures. Everyone needs access to the same information and a respect for the impact they can have when calling the shots. Testing tools give people access to that data. But they all have slightly different ways of presenting results. Underpinning those visualizations are foundations they all rely on. When you understand the foundations, you can read the results in any tool. The foundations are:

- **Sampling distribution.** When you run an experiment, you observe one result, but you could have observed many others. The sampling

---

6. SRM is short for *sample ratio mismatch*. In case you forget from Chapter 3 in the section "Randomization & Sample Ratio Mismatches (SRMs)" it simply means that your sample group isn't split according to the intended percentages you chose up front. Most of the time it's a 50/50 split.
7. This book focuses on explaining simple 50/50 split, frequentist, fixed-horizon A/B tests. This is intentional.
8. Some experimentation platforms help you decide the end date and time and stop it for you automatically.

distribution tells you how likely it is to observe each possible result. This is a fundamental element in experiment analysis.

- **P-value (short-hand for probability value).** This is the probability of observing data as extreme—or more extreme—than what you've already seen in your dataset. It is assuming that the change you made has no impact. (This refers to your null hypothesis.)[9]
- **Confidence interval and level.** A confidence interval is a range around an estimated value. It's used together with a confidence level to quantify the uncertainty around the estimate. The range gives you information about the *real* value (which is impossible to know), while the confidence level tells you how certain you can be about the estimated value.

Now, it's time to look into your data crystal ball: your experiment tool! It's time to predict the unknown.[10]

Calculating the Conversion Rate

Huzzah! Your experiment just finished running. Now, you've got some data. Your testing tool has captured two very important numbers for both the control and the treatment in your test:

1. **The number of experiment participants.** This is your sample size.
2. **A count of how many of those experiment participants "converted."** For example, how many people completed a purchase or called customer service.

With these two numbers, your tool can calculate conversion rates. To do this, your tool assigns a "conversion status" to each visitor. Visitors who converted are assigned a one, and users who did not convert are assigned a zero. The tool uses this data to calculate the average conversion status for all experiment participants. It does this by dividing the number of conversions by the number of visitors. Then it turns that average into a percentage by

---

9. It's OK if you read this many, *many* times and still don't understand what it means. P-values are notoriously the most difficult concept in experimentation. If you're starting to have a slight panic attack after reading this, just remember: You don't have to understand all these concepts perfectly to make better decisions while using them.
10. Typically fortune tellers are depicted as people in flowing robes. Turns out, the most qualified people to make predictions are just nerds with very strong computers.

multiplying by 100%. This process is the formula for calculating a conversion rate, shown in Figure 7.2.

Number of Conversions ÷ Number of Visitors × 100% = Conversion Rate %.

FIGURE 7.2 This is how to calculate a conversion rate.

Once you have the conversion rates, you need to visualize the result's distribution. Many types of distributions exist, and they're easily visualized. These visualizations help you understand how confident you can be in the reported mean. But before you dig into data visualizations, it helps to go through a thought exercise to understand how statistics work.[11]

Imagining "Infinity" Experiments and Beyond!

When you run a conversion experiment, for each variant (control or treatment) your tool computes *one* conversion rate. Imagine you could run the exact experiment again on a new sample of users of the same size (same dates, same weather, same everything—just different users).[12] If you did that, you'd very likely observe a different conversion rate than you did for the first run. Now, imagine you were to run the same experiment a third time with a new sample. You'd likely observe yet another different conversion rate. And now, imagine that you ran the same experiment with new samples of the same size an *infinite* number of times. After an infinite number of runs of the same experiment, you'd have a *ton* of conversion rates, each tracing back to a single, unique run. But because there are so many conversion rates, you'd likely observe *many* different values. And each of these different conversion rate values would all show up in your data at different frequencies.

---

11. The next section is for hardcore experimentation nerds. If you're up for it, strap in and give yourself time to digest this stuff. If you're not, that's also fine. I drew Figure 7.7 as a cheat sheet, so you can also just skip ahead to the section "Looking at the Dots and Lines" if you prefer.
12. This concept is kinda like thinking about parallel universes.

A *frequency* is how often something happens. In this case, the frequency is how many times a specific conversion rate is observed when you run an infinite number of the same experiment. To get analyzable data, you need to visualize how many times each conversion rate shows up in your data set. If you did that, in this case, you'd get a visualization similar to what statisticians call a *sampling distribution,* as illustrated in Figure 7.3.

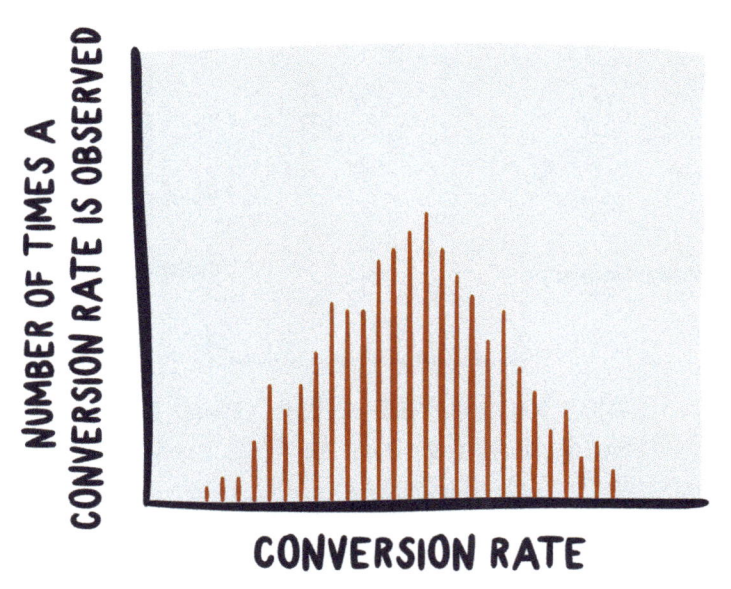

FIGURE 7.3 A simple visualization of how many times a value shows up in a data set.

The sampling distribution assigns a probability to each possible result of every run of your infinite number of experiments. A *probability* is a number that represents how likely something is to happen. The probability assigned to each value in the sampling distribution helps you understand how likely it is for you to observe a conversion rate if you run the same experiment again on a new sample of the same size.

For example, suppose that you run an experiment with 100 users in the treatment. At the end, you observe a conversion rate of 10%. Now, imagine that you somehow know the *real and exact* sampling distribution for your experiment. (This is impossible to know.) The actual sampling distribution tells you that even though you observed a 10% conversion rate, you could have just as easily observed *any* value in the range from 5%–15% with a probability of 80%. This is illustrated in Figure 7.4.

SAMPLE SIZE: 100
The conversion rate you observed.

10%
5%   15%

80% chance the value could have landed anywhere in this range.

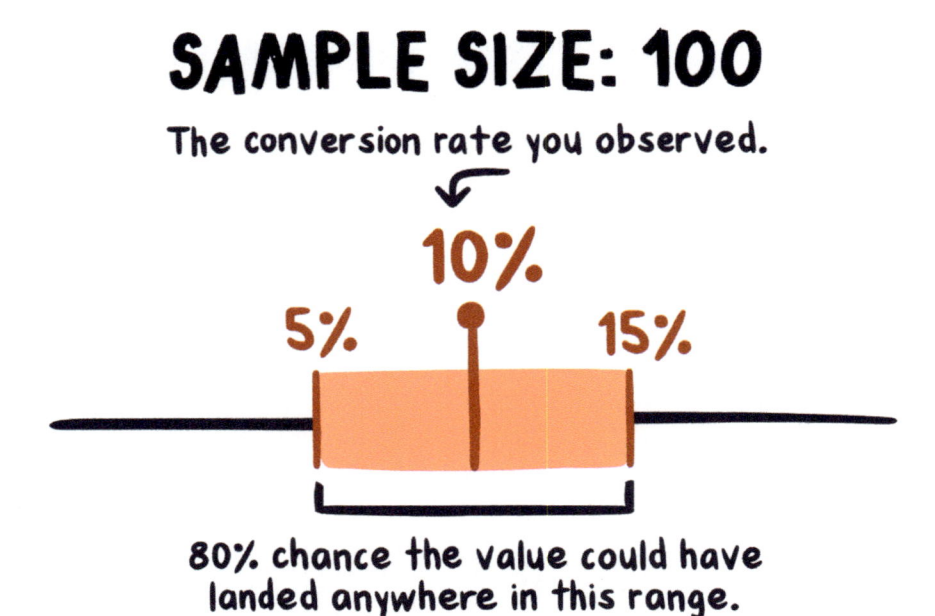

FIGURE 7.4 Here is the conversion rate you observed and the range where the conversion rate is likely to fall should you run the experiment again.

The information in the sampling distribution tells you that the conversion rate could vary a lot. This suggests that you probably shouldn't put a lot of confidence in the 10% you observed.

To understand why this is important, imagine that you run this same experiment again, but with 10,000,000 users in the treatment this time. When the experiment is over, you observe a conversion rate of 11.99%. Again, imagine that you somehow know the *exact* sampling distribution of that experiment run. The sampling distribution tells you that even though you observed 11.99% conversion rate, you could just as easily have observed any value in the range from 11.98%–12.00% with a probability of 80%. This is illustrated in Figure 7.5.

SAMPLE SIZE: 10,000,000

The conversion rate you observed.

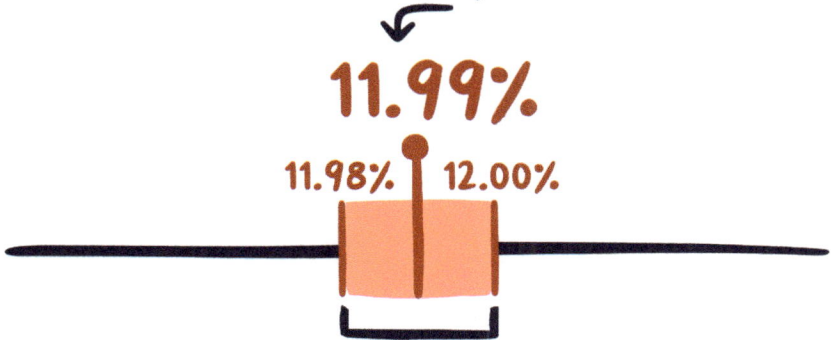

11.99%

11.98% | 12.00%

80% chance the value could have landed anywhere in this range.

FIGURE 7.5 The range is much smaller this time.

Based on the sampling distribution, you can see that the conversion rate isn't varying a lot this time. 80% of those imaginary conversion rates would fall between 11.98% and 12.00%, which is a much smaller range than the previous 5%–15%. This suggests that you can have more confidence in the 11.99% you just observed than in the 10% you observed in the sample of only 100 people.

So, what's the point? Even though it's impossible to get the actual sampling distributions for an experiment run, you can still approximate one. And, luckily, because a conversion rate is an average, you know that the sampling distribution will have a very specific shape. And that shape is called a *bell curve*.

A BRIEF HISTORY OF THE BELL CURVE

The most well-known type of distribution is called a *bell curve*. But it's also known as *normal distribution*, a *Gaussian curve*, or a *Laplacian curve*—it's recognizable because of its symmetrical shape. Dr. Andrew Vickers, a urologist and world-renowned cancer researcher, explains normal distribution in his book, *What Is a P-Value Anyway?*: "When you add up a lot of chance events you get a normal distribution."

When I read in Dr. Vickers's book that "when you add up a lot of chance events you get a normal distribution," I screamed: "HOW IN THE WORLD DID SOMEONE FIGURE THAT OUT?!"[13] Naturally, I Googled it. Abraham de Moivre is credited with discovering normal distribution in 1733. He was a gambler who enjoyed a dice game. Over time, he kept track of all the rolls. I assume he did this to create a strategy to improve his earnings. Eventually, he realized that more or less each time he calculated the probabilities of the rolls in the games, the curve looked similar. To see if it was a fluke, or if something else was going on, he took other types of measurements to see their distributions. Amazingly, he found the same phenomenon—over and over and over again. Normal distribution can be found in measurements of human heights and weights, to daily average temperatures, to the height of trees. It has many practical applications, one of which is A/B testing.

Although Moivre discovered normal distribution, it was German mathematician Carl Friedrich Gauss who made it famous. He put normal distribution to use in astronomy and blew people's minds by predicting the orbit of the dwarf planet, Ceres. So, everyone started calling it a "Gaussian curve." Gauss wasn't the only one who expanded on Moivre's work, though. Pierre-Simon Laplace, a French scholar, took Gauss's work even further. He used normal distribution to predict other complex things such as population growth. We still use these principles that Moivre, Gauss, and Laplace discovered and developed to make sense of the world around us—and to get better at gambling!

---

13. Math makes me shouty.

Ringing in the Bell Curves

A bell curve (illustrated in Figure 7.6) can be drawn with just two numbers: its mean and its variance. This is similar to how a rectangle can also be drawn with just two numbers—its height and its width.

Bell curves are important because they're a common experiment result visualization. These graphics are a concrete artifact to understand how confident you can be in the conversion rate you observe. If you knew the real conversion rate and variance of your population, you could easily just draw the corresponding bell curve. But even though you don't know these exact numbers, you can reliably approximate the bell curve anyway. To do this, you use the sample size and the observed conversion rate in a single experiment run. Amazingly, these numbers also contain the data set's approximate variance, which you can use to draw your bell curve.

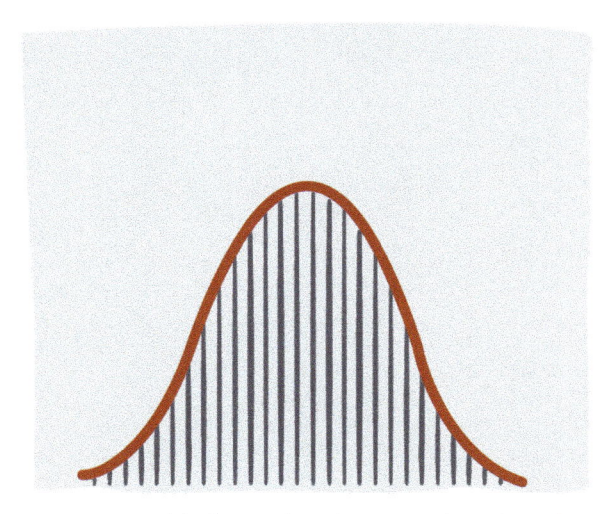

FIGURE 7.6 A bell curve is an important shape in math.

At this point you might be thinking: "WHY DID I JUST SUFFER THROUGH NONSENSE ABOUT ONES AND ZEROES, AN INFINITE NUMBER OF IMAGINARY EXPERIMENTS, AND HOW DRAWING CURVES IS LIKE DRAWING RECTANGLES?"

And the answer to your question is: Because taking the conversion rate data and drawing a curve with it is precisely what you need to analyze your experiments. Figure 7.7 illustrates the concept of sampling distribution.

The Sampling Distribution

The sampling distribution is a statistical concept. It helps you understand if your data contains enough evidence to reject your null hypothesis or not.

The Gist of the Sampling Distribution:

Step 1.
You want to learn the conversion rate for an entire population. Some people in the population converted, others did not.

Step 2.
You take random samples of the same size from that population. In this example, it's 10 random samples of 5 people each.

Step 3.
You calculate the corresponding conversion rate for each sample. This number is an average. It's *very important* that a conversion rate is an average. Because although an average is a single number, it contains *a lot* of useful information.

Step 4.
You count how many times each conversion rate is observed and put that data into a table. This helps you learn which conversion rate(s) show up the most frequently. The frequency of an observed conversion rate helps you understand the probability of that number being observed again if you ran the same experiment many, many times.

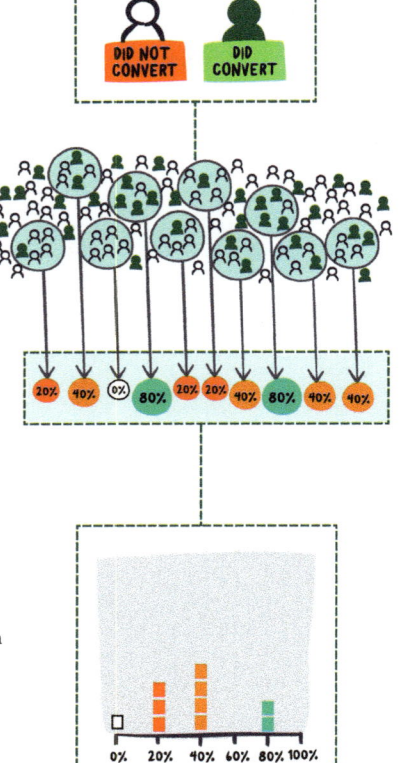

FIGURE 7.7 The sampling distribution is a core concept in statistics.

Step 5.
Do steps 2–4 many, many times. Eventually, as the data stacks up, a bell curve forms.
This curve is the sampling distribution of the conversion rate for the population of people you want to learn about. To draw a bell curve you only need two numbers: the conversion rate and the sample size. In this case, the sample size is *very* small. So, the data doesn't contain enough evidence to give you a good indication of the conversion rate.

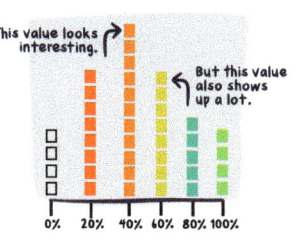

Step 6.
To learn the conversion rate for your population of interest, look at the shape of the bell curve.
Because the sample size was so small—only five people—the curve is pretty wide and slopes gradually. This means that there's a lot of variance in the data. The conversion rate could be 40%, but it's also highly likely to be any value between 20% and 60%.

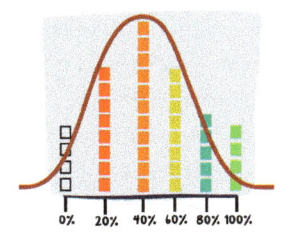

But wait—there's more!

To get more reliable results, you can increase your sample size.

Imagine you do steps 2–5 again, except you use a sample size with lots of people—say 100,000 people instead of 5. Then you get a much sharper curve. This happens because the larger the sample size, the more concentrated the sampling distribution is around the population's mean.

This is good because the sharper the curve, the more you can rely on a single sample to learn about the conversion rate for your population of interest.

So what now? You can build confidence in your data.

The sharper the curve, the more you can rely on a single sample.

Even though this concept is a thought exercise, the sampling distribution for a population can be approximated from a single sample. Then you can use it to construct a confidence interval. A confidence interval is a range of values. And in experimentation, a confidence interval allows you to decide if the data contains enough evidence to reject the null hypothesis or not.

Analyzing Experiments

In null hypothesis testing, analyzing experiment data means to decide if you have enough evidence to reject the null hypothesis or not. Here's one way to do that:

First, using the experiment data (which is just the conversion rate and the sample size of each variant), approximate the sampling distribution of each variant. You do this by drawing the corresponding bell curves side-by-side.

Next, you want to compare the conversion rates. But you don't want to compare the exact observed values. That's because you know that you easily could have observed a different value. There's actually nothing special about the specific one you observed—it just got lucky. Making decisions based on lucky numbers is not reliable at all. Instead, you want to compare a range of values, which contains many, many possible experiment results.[14] This approach reduces the influence of that single observed value on your decisions, which gives you more confidence. Something to note now, is that a different word for *range* is *interval*. Hence, a *range* of possible results if you would run an experiment again and again, is called a *confidence interval*.

A *confidence interval* is a range of values that covers a desired area of the sampling distribution. For example, we can say with 100% confidence that the observed conversion rate of a variant is in the range 0% to 100%. That's because 100% of the possible conversion rates of a variant are between 0% and 100%. This is a 100% confidence interval, and it's not very useful. Why? Because, sure it gives you 100% confidence, but it also gives you *no information at all*. To remedy this, you need to give up some confidence to get some information.

In the first example previously (Figure 7.4), you learned that 80% of the possible results of the experiment are in the interval between 5% to 15%. This means it's the 80% confidence interval. In the second example (Figure 7.5), the 80% confidence interval was 11.98% to 12.00%. Both intervals have the same confidence level: 80%. But the first one is much wider than the second one, which means it gives you much less information. When you decide that the confidence level is 80%, you accept that with a probability of 20%,

---

14. Do you understand why you needed to imagine an infinite number of experiments now? Reliability comes from imagining how consistent the outcome of the experiment is should you run it again.

the experiment results could be out of the reported interval. That's just the uncertainty you have to accept if you want to get more information. **That's why it's useful to think of a confidence interval as a tool to buy information, by paying with uncertainty.**

A confidence interval is not an interval for the real conversion rate. This is a common misconception. A confidence interval is a range for the *potential results* of an experiment. It helps to quantify how likely it is that you could have observed a different result to the one you actually did. A wide confidence interval tells you that even though you observed a specific conversion rate—you could have very well observed other very different values, too.

GUESS CONFIDENTLY THE NUMBER OF CANDIES IN THE JAR

When I was little, I'd go to my Aunt Erna's house for family holidays. Her "guess-how-many-candies-in-the-jar" game was always a hit. She'd fill a jar with a number of candies that she counted in advance. Each player would guess the number of candies and write it on a piece of paper. Whoever came closest to the true number won the jar of sweet rewards.

The real fun (and my personal favorite strategy) was to walk around before guessing to hear others talk about their guesses. "Oh, the Skittles?" said my Uncle Big. "I guessed there were 587." "Shows what you know!" retorts my dad. "I'm 95% certain there's no more than about 300 in there. It's gotta be somewhere between 175–300. That's why I guessed 225." After hearing that, Big also wants to be 95% confident because he can't possibly let his older brother one-up him. But he's less sure, so he says, "Well, I'm 95% sure it's somewhere between 200–900!"

At this point, it's clear that even though they're both 95% confident, their ranges vary wildly. Without realizing it, my Aunt Erna and my arguing dad and uncle were schooling me in a masterclass about confidence. Anyway…at the end of this particular game, Erna declares that the jar has 283 candies. And in a jarring twist—Cousin Greg wins.

This means that you shouldn't rely on a single observed value too much. A narrow confidence interval tells you that with some probability (given by the confidence level you chose), you would observe very similar values. This makes the observed conversion rate a good representative of all the potential conversion rates that were not observed. To compute an approximate confidence interval for conversion rate, you need:

- **The observed conversion rate** (e.g., 6%), or alternatively the number of converted users (e.g., 20,700)
- **The sample size** (e.g., 345,000)
- **The desired confidence level** (e.g., 90%)

Then you enter those three numbers into a confidence interval calculator to get two new numbers:

1. **The lower bound** of the interval (e.g., 5.93%)
2. **The upper bound** of the interval (e.g., 6.07%)

Now you have one confidence interval for treatment and one for control. With these numbers you want to answer the following question:

> **Do you have enough evidence to reject the null hypothesis of no difference between control and treatment?**

Comparing the confidence intervals helps you answer that question. If confidence intervals do not overlap, you can be sure you have enough evidence to reject the null hypothesis. A reasonable conclusion from this result is that treatment and control have different conversion rates. But, what if they do overlap? In principle, this suggests that you don't have enough evidence to reject the null hypothesis. This conclusion is likely correct. However, in some cases a small overlap will still allow you to reject the null hypothesis. To be sure you're drawing the right conclusion, always check if the reported p-value is bigger than alpha.

Another popular way to answer the question, "Do you have enough evidence to reject the null hypothesis?" is to compute a confidence interval for the difference between the conversion rates of the treatment and the control. You get this number by subtracting the conversion rate of the treatment minus the conversion rate of the control. With this approach, you use a different calculator. In it, you must enter the sample size and conversion rates of each variant and the confidence level (given by 1-alpha), and then the calculator gives you one single confidence interval. You can now simply check if the interval contains zero. If it doesn't, it means that if you were to repeat the experiment many times, it is very unlikely that you will see no difference. This suggests that there is indeed a difference between the conversion rates of the control and the treatment. This means, you should reject your null hypothesis. If the confidence interval does contain zero, it means that there is some probability to observe zero or very close to zero values. This suggests that you don't have enough data to reject the null hypothesis of no difference in conversion rate between the control and the treatment. This approach will always agree with the p-value check. But you should always check if the p-value is less than alpha, and only then, settle on a conclusion.

Probably the Most Confusing Part of the Book: P-values

P-value is short for *probability value*. It's the number you worked *so hard* to know. It's what tells you if you should accept or reject your null hypothesis. P-values are calculated using (yet another) fancy formula and a bunch of numbers from your experiment data. The p-value and your alpha play a big role in the final analysis of your experiment. You chose your "alpha" when you calculated your sample size. (Check out Chapter 6 in the section called "I've Got the Power [and Significance]" if you forgot already.) Boiled down, the main question you need to answer in your analysis is:

> **Is the p-value the same or smaller than the alpha (significance level) used to calculate the sample size?**

If it is, then the p-value has reached statistical significance. You should reject your null hypothesis and accept the alternate hypothesis as the new default belief. This should reflect what you captured in your Good Experimental Design toolkit in the Validate template. If it's not, then the result is not statistically significant. You should not reject your null hypothesis because you didn't get strong enough evidence to sway you from your default belief. In other words, you won't change your mind. Figure 7.8 summarizes it for you.

> **Significant**
> P-value ≤ Alpha
> (less than or equal to)

> **Insignificant**
> P-value > Alpha
> (greater than)

FIGURE 7.8 The p-value reported in your result determines if the result is significant.

You want to learn if the result is not simply by chance. The reported p-value helps you learn that.

Confidence and Statistical Significance

The confidence reported in your experiment result is calculated with the p-value. To calculate the confidence, you simply subtract the p-value from the number one and convert the result to a percentage. Conversely, you can turn the p-value into a percentage first and then subtract it from 100%—whatever breaks your brain less.

Imagine that you ran an experiment where you chose an alpha (significance level) of .1 (10%). You used your alpha to calculate your sample size when you filled out the Metrics & Math portion of the Validate template in the Good Experimental Design toolkit. (This means that there's a 10% chance a test with no *real* effect could report a false positive.) This translates into needing a confidence level of 90% to consider the result statistically significant. A result is considered statistically significant if it reaches the significance of .1 or smaller. (This is the same thing as 90% confidence or more.) You decided this number *before* you ran your experiment as the threshold at which you'd "change your mind" and reject the null hypothesis.

When your experiment reaches the necessary sample size and calculated run time, it's ready to analyze. To analyze the result, you look at the p-value. The p-value is .06. Again, the p-value is directly related to your confidence. To get the confidence from a p-value, you subtract the p-value from the number one and convert it to a percentage. Or you can turn the p-value into a percentage and then subtract it from 100% In this example, it works out to be 94%, as shown in Figure 7.9.

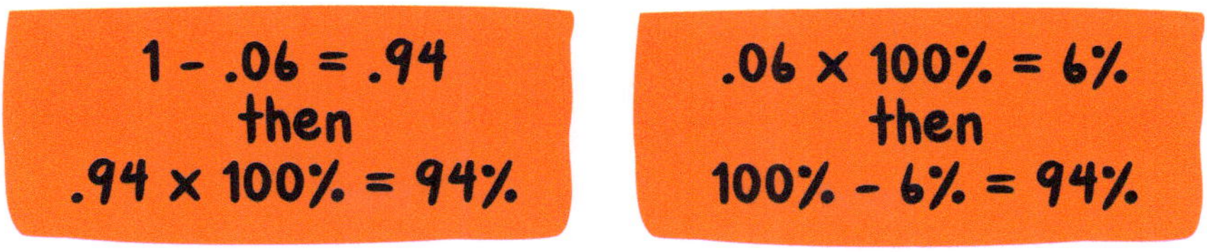

$$1 - .06 = .94$$
then
$$.94 \times 100\% = 94\%$$

$$.06 \times 100\% = 6\%$$
then
$$100\% - 6\% = 94\%$$

FIGURE 7.9 Two ways of calculating the confidence from a p-value.

To know if the result is statistically significant, you must answer this question:

> "Is the p-value the same or smaller than the .1 (significance level) used to calculate the sample size?"

Conversely, the question could also be asked in this way:

> "Is the confidence higher than 90% based on the reported p-value?"

The answer to this question in the example is "yes." Why? Because .06 is less than .1, and 94% is more than 90%. Therefore, the result is statistically significant.

That being said, some experimentation tools and platforms report only p-values, others report only confidence, and others report both. As long as you have either a p-value or its corresponding confidence, you can determine if a result is statistically significant or not.

Looking at the Dots and Lines

When you reach your decision moment, you'll see a bunch of stuff on a dashboard. This usually includes some numbers in a table and possibly some "graphy" looking things (Figure 7.10).

These fancy visualizations simply compare the means between base (aka the *baseline*) and variant. Other numbers you need to make sense of the results, even if they're not specifically labeled, are:

- **Visitors.** This is the number of people in base and variant. This helps you know if the split is correct and if you've reached your sample size.
- **Conversions.** The number of visitors who completed the goal in both base and variant.

- **Conversion rates.** The conversion rate per variant.
- **P-value.** The number that determines if the result is statistically significant or not based on your significance level.

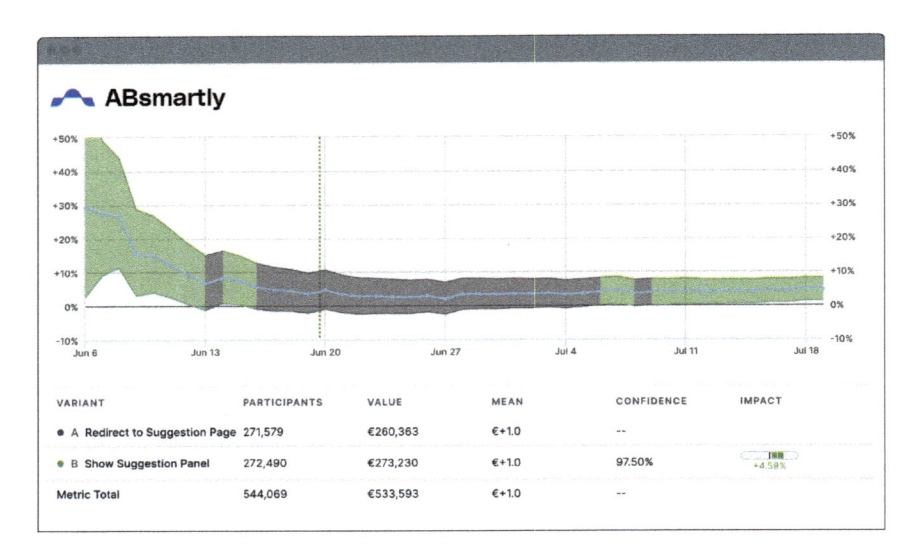

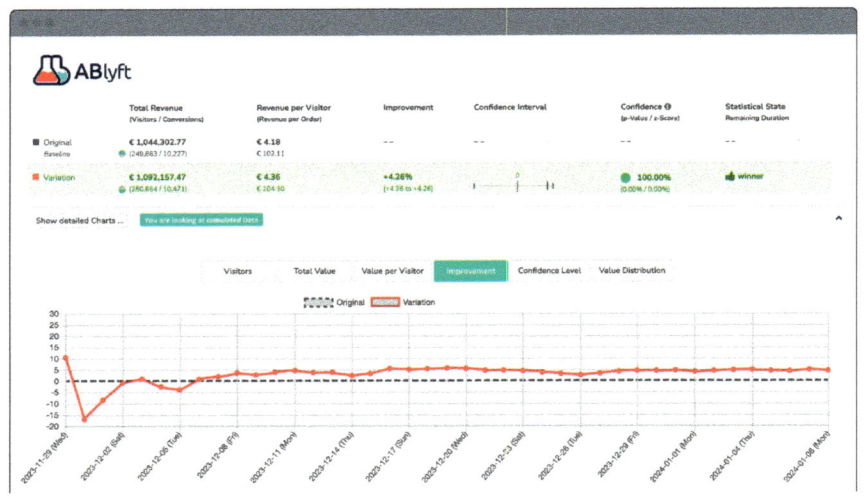

FIGURE 7.10 Screenshots of how experimentation platforms ABsmartly (top) and ABlyft (bottom) visualize experiment results.

Checking the Result

Once you've located all the numbers you need, do the following things to analyze the result:

1. Check the number of visitors and the split between base and variant. These numbers answer these questions:

    - **Did the test reach the sample size it needs to give valid results?** The right sample size needs to be reached to consider the result valid. If it doesn't reach the sample size, you might hear someone say the result is *underpowered*. Make sure that the test is at least the number that you calculated in the Test phase and wrote in your Validate template.

    - **Is the distribution correct?** If you ran a 50/50 split test, you should have about 50% of the overall visitors in base and the other 50% in the variant. You can check with an SRM calculator to make sure that it's correct.

    - **If the answer is yes to both questions, then the result is most likely reliable.** Ideally, you consider your business cycle, so your visitor behavior is balanced, and the results aren't biased. Figure 7.11 shows how and where the number of visitors is shown in both ABsmartly and ABlyft.

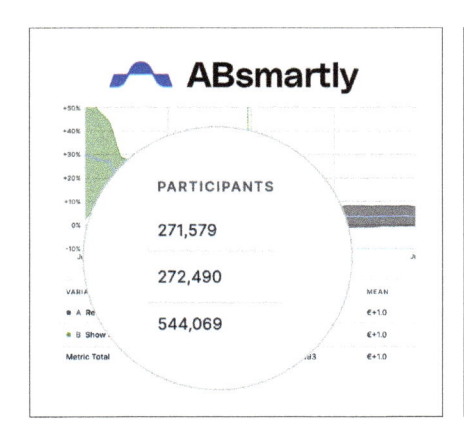

FIGURE 7.11 (Left) How ABsmartly shows the number of visitors (aka participants). If there's no error message in a tool that does automatic SRM checking, the split is correct. (Right) ABlyft shows the number of visitors tucked away under the Total Revenue heading.

THE ANALYZE PHASE: LEARN FROM THE DATA · 191

2. Check the conversion rate and the confidence for base and variant. These numbers answer these questions:

- **Is the conversion rate in base or variant higher than the other?** You need to understand if there's a difference between the means in base and variant.

- **If so, is the difference statistically significant?** You look at the p-value and/or the confidence (shown in Figure 7.12) to see if any difference between the two is unlikely due to chance. If it's unlikely due to chance based on the significance level you chose before you ran the test, then you can reasonably assume it's a real effect.

FIGURE 7.12 How ABsmartly (left) and ABlyft (right) visualize experiment results.

> **TIP** DON'T MAKE A DECISION BY COMPARING THE MEANS AND IGNORING THE REST
>
> Some people simply compare the means between base and variant and declare a "winner" if one is higher than the other—even when a result is insignificant. This is a problem because you must consider the p-value and confidence intervals around the means, too. For example, if you had a result that showed the variant increased conversion by 1.2% (±2%), that's just the "best" guess.
>
> It's a pretty big gamble to make decisions with inconclusive data. Of course, there's nothing technically stopping you from doing this. But you need to be honest with the fact that you're simply loosely hedging your bets and not confidently making a product improvement. You should also report the level of uncertainty of the result for the sake of personal integrity and intellectual honesty. ■

Visualizing the Results

As mentioned earlier in this chapter in the section called "Confidence and Statistical Significance," confidence and p-value are basically two different presentations of the same thing. Tools typically show confidence as a percent, and/or the p-value as either a percent or decimal. Don't get worried if you see decimals because you can always convert them to a percentage. When you look at the results in Figure 7.13, you see that the p-value is 0.03%.

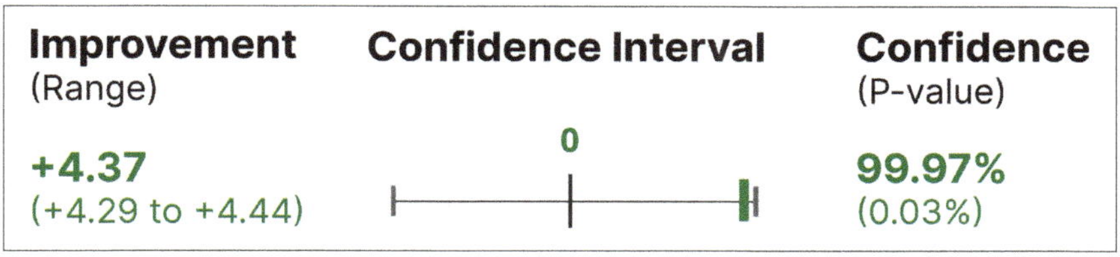

Improvement (Range)	Confidence Interval	Confidence (P-value)
+4.37 (+4.29 to +4.44)	⊢———┼———⊣ 0	**99.97%** (0.03%)

FIGURE 7.13 This shows a p-value of 0.03%, which is equal to a confidence of 99.97%.

The 0.03% converts into the confidence simply by subtracting it from 100%, as shown in Figure 7.14.

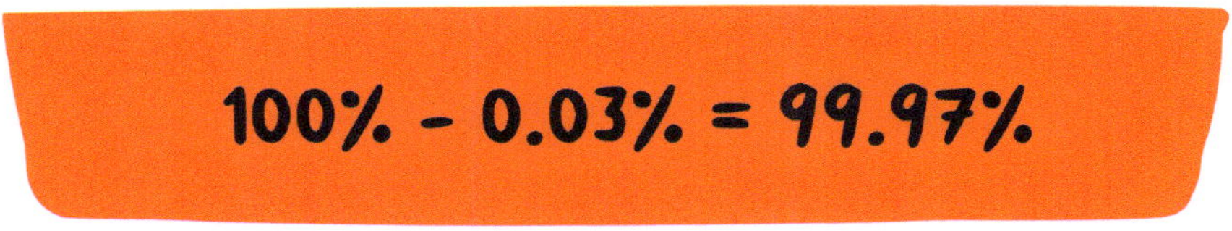

$$100\% - 0.03\% = 99.97\%$$

FIGURE 7.14 In a tool, you'll either have a p-value, confidence, or both to learn if a result is statistically significant.

BOX PLOTS

Your confidence is sometimes visualized in a *box plot*, shown in Figure 7.15.

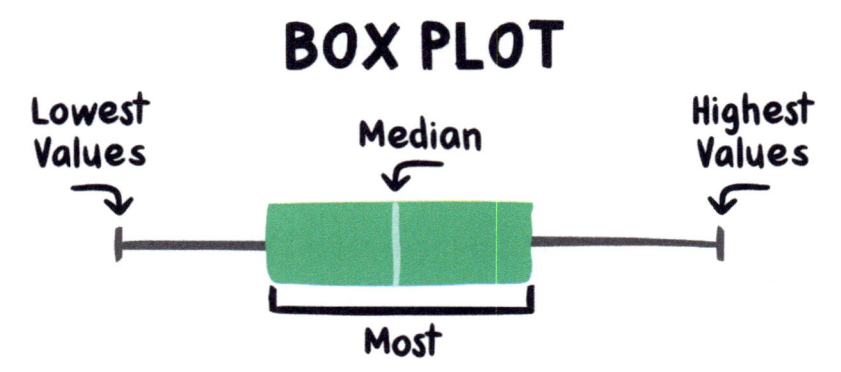

FIGURE 7.15 Many experimentation platforms visualize test results with a box plot.

The confidence visualization can vary wildly. But they all show the same things: the means for the baseline and variant. Figure 7.16 shows two different box plot visualizations.

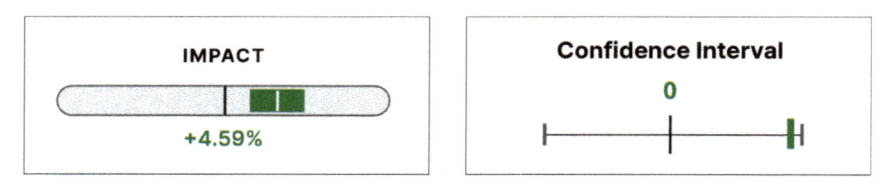

FIGURE 7.16 These visualizations are styled differently, but they represent the same things.

LINE CHARTS

Another type of visualization is a line chart. Line charts show the baseline and how the variant's mean compares to it over time. As with the box plot, some line charts show the interval around the mean and others don't. The data for the base is always shown in the constant line at zero. This baseline represents the current state and how the variant compares. You'll likely see the lines land above and below the baseline, which shows where the variant's mean and the range around it lands over time. When the mean and the likely range around it hover around the baseline with overlap, the result is insignificant. When the mean and the range around it consistently land above or below the baseline, the result is significant, as shown in Figure 7.17.

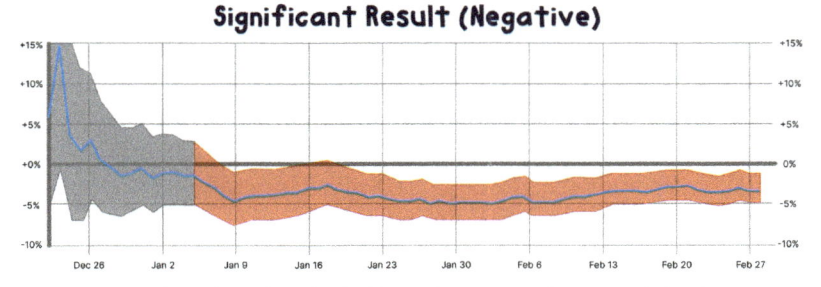

FIGURE 7.17 What a line chart looks like with an insignificant result (top), a positive result (middle), and a negative result (bottom) with a range around the mean.

PLUS-MINUS SYMBOLS

Finally, you may see something that looks like this ±. It's a plus-minus symbol. This symbol clarifies the range of possibilities around the most likely mean. It typically includes two standard deviations around the mean, which usually accounts for 95% confidence.[15] To sum it all up, all tools will show you some kind of visualization of how the mean and the range around it compare to base, as well as the level of confidence you can have of the result.

---

15. This is what many people use. However, it will reflect whatever you choose for your significance level.

EXPERT ADVICE FROM THE FIELD

INTERACTION EFFECTS—DON'T WORRY ABOUT THEM! WITH LUKAS VERMEER

I've given dozens of talks about experimentation statistics across the globe. During the Q&A, I always get asked the same question: "But, Lukas...What about experiment interaction effects?" People believe that if you run more than one test at a time, they influence one another in ways that make the results unreliable. Although this is possible, it's highly unlikely.[16] My overwhelming answer to this question is: "Don't worry about them! You worrying about potential interaction effects is a bigger problem than any miniscule chance of one actually happening."[17]

Now that you know you shouldn't worry about interaction effects, let's dig into why. First thing to know is that the amount an interaction could happen depends on the type of interaction it is. Interactions come in two flavors: metric interactions and traffic interactions.

Metric interactions are when two experiments combine to have a measurable impact on an important metric. For whatever reason, the sum total of the two things together is different than when looked at individually. A metric interaction typically happens because of something called a *functional conflict*. (In other words: something breaks.) It can also be when something unexpected happens (in either a good or a bad way) when the two experiments combine. Bugs are the most likely scenario. "Happy accidents," the other scenario, are when two experiments combine to have a positive effect. These are much less likely. An example of a functional conflict interaction is this: Imagine one team experiments with turning the background of a button pink, while another experiments with turning the text on a button pink. Separately, they work just fine, as shown in Figure 7.18.

---

16. Monwhea Jeng, "A/B Interactions: A Call to Relax," *Experimentation Platform* (blog), August 2, 2023, https://www.microsoft.com/en-us/research/group/experimentation-platform-exp/articles/a-b-interactions-a-call-to-relax/ and Eric Haavind-Berman, "How Carwow Accelerates Online Experiments with Overlapping AB Tests," *Medium*, December 4, 2023, https://analyticsatcarwow.medium.com/how-carwow-accelerates-online-experimentation-with-overlapping-ab-tests-a59a918fc6d2
17. In the source article, Microsoft found only a small number of potential interactions: about 1 in 50,000 A/B test pair metrics.

The hypotheses behind the individual changes are sound. But, when each experiment is shown at the same time—something unintentional happens: pink text on a pink background. What would otherwise be two fine ideas come together to create a problem, as shown in Figure 7.19.

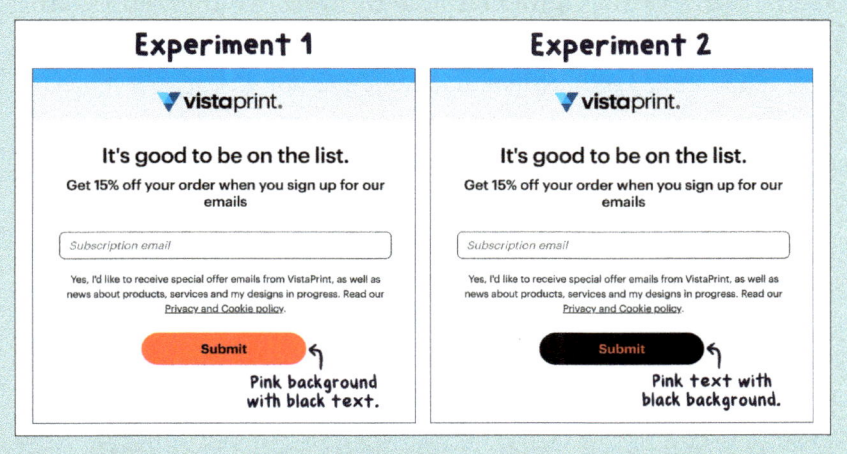

FIGURE 7.18 Two experiments separately that cause no problem. One test changes the text color in a button pink, and the other test changes the background color of the button to pink.

FIGURE 7.19 Pink text on a pink background is an example of a functional conflict.

CONTINUES ➤

EXPERT ADVICE FROM THE FIELD

CONTINUED ➤

This type of functional conflict is easily prevented with quality assurance (QA) testing. Most tools allow you to see products with all experiments off and all experiments on, which helps you catch issues like this. Then you can avoid potential functional conflicts that might cause an interaction effect.

To solve an identified functional conflict, you could either…

- **Split the traffic.** This solution is often not practical for most products because most businesses have too little traffic to get the sample size they need for reliable results in a decent amount of time.

- **Coordinate the experiments to run them at different times.** If the experiments aren't live at the same time, then they have no opportunity to conflict. This means one team will have to wait.

As long as there are no functional conflicts, you can run as many experiments as you want in the same area of a product.

Traffic interactions happen when one experiment changes the flow of users to another experiment. It's like changing the course of a river and watching how it affects the landscape downstream. This doesn't usually cause problems for users or decision-making. But it can introduce some bias into the results.

For example, imagine a landing page that has a call-to-action button experiment. In the control, the button goes to a search results page, and in the treatment, the button goes to a product page. The traffic to experiments running "downstream" from this experiment is now different. That's because the "upstream" experiment forces people through a different path. If either the control or treatment "upstream" is a much better or much worse experience, then the results of the "downstream" experiment might get thrown off. More people might get through the better experience, which redirects traffic to another experiment, as shown in Figure 7.20.

Although this might seem like a problem, it's highly unlikely that traffic redirection will skew the results enough to change the outcome. Lots of research has been done about this to understand the likelihood of it happening. The overwhelming consensus within the data science community is that these types of interactions rarely happen. Even when they do happen, the impact is typically not big enough to cause concern. And when the effect is big enough, they're often easy to spot.

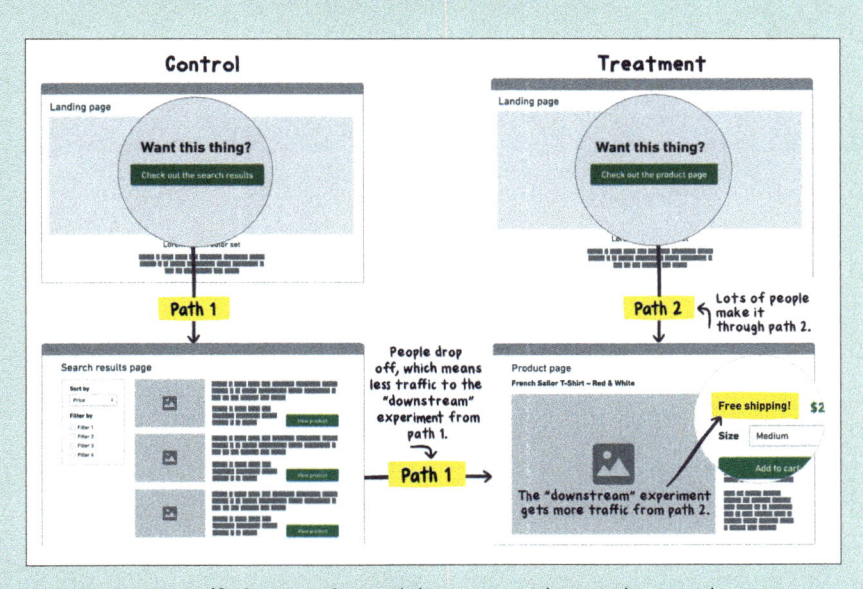

FIGURE 7.20 Traffic being redirected down one path more than another.

CONTINUES ➤

EXPERT ADVICE FROM THE FIELD

CONTINUED ➤

So, again—don't worry about interaction effects. If you check for and avoid functional conflicts, you'll catch many of the potential bugs that may cause a metric interaction. And traffic interactions are so rare that they're not worth even looking for. Just go run more tests!

Lukas Vermeer
Head of Experimentation at Vista

Lukas's specialty is designing and building the infrastructure and processes required to start and scale A/B testing to drive business growth. You can learn more about him, use the tools he makes, and read the stuff he writes at www.lukasvermeer.nl.

The Important Bits

- You can "look," but don't "peek." Otherwise, you increase your chance of observing a false positive.
- Build data intuition by looking at many experiments over time.
- Data can be very volatile when the sample size is low, so learn how data behaves to get a feel for when something is actually wrong.
- Math (specifically statistics by way of calculating probabilities) helps you understand if the change you made had any impact. You compare the conversion rates of the control and the treatment to see if they're significantly different. (You choose what you consider is "significant" in your sample size calculation when you select your alpha.)
- Experimentation platforms have different ways of visualizing and reporting results. The main things you need to look for once your experiment run time is up, are the sample size (make sure that you have enough participants and that the sample is split correctly), p-value or confidence, and the reported conversion rate with the confidence interval.
- You can reject the null hypothesis if your experiment reports a statistically significant result. Otherwise, the data aren't reliable enough to draw confident conclusions.
- Don't worry about interaction effects. It's OK to run many tests in the same product, in the same area at the same time, as long as two experiments combined don't cause any bugs. Do good quality assurance testing before starting your experiment to prevent functional conflicts from happening.

CONVERSION DESIGN PROCESS

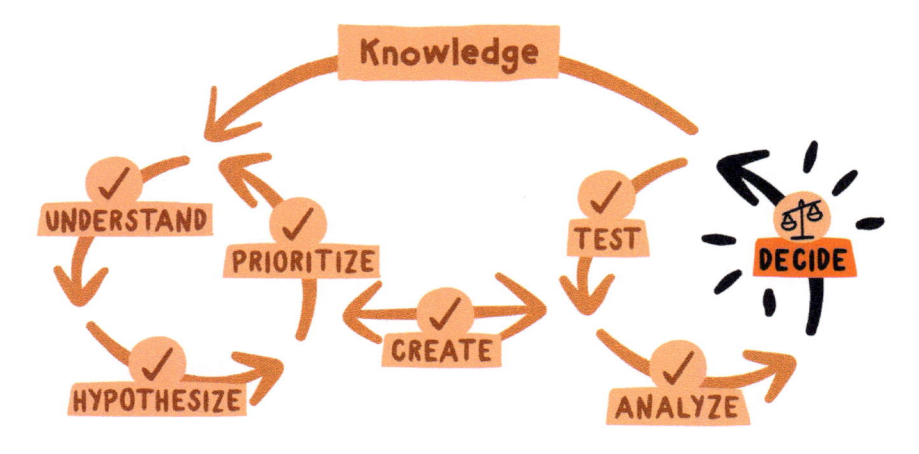

Knowledge → Understand → Hypothesize → Prioritize → Create → Test → Analyze → Decide

"You cannot make progress without making decisions."
— Jim Rohn

CHAPTER 8

The Decide Phase: Make an Optimal Choice

THE POINT

Use your experiment analysis to understand the short- and long-term impact of your change. Then make an optimal decision to create as much value as possible in both the short- and long-term.

My dad is a very funny, cynical person. He's known for his general suspicion of people—especially in large groups. "Whenever there's a problem, you can bet there's always at least one person involved," is something he's said more than once. And, honestly, he's right.[1] For some reason, people—especially in large groups—do stupid things and make stupid choices.

---

1. Hi, Papa. Thanks for editing this book for me. Sorry I called you cynical. But I also called you right!

Karl Albrecht, an organizational consultant, and author of *The Power of Minds at Work*, coined this concept as "law," which holds that "intelligent people, when assembled into an organization, will tend toward collective stupidity." Conversion Design, with its focus on learning, exists to help teams and companies make better decisions consistently and at scale to avoid collective stupidity that is common in companies. Healthy skepticism, regular reflection, and intellectual honesty are key to keeping people (individuals or in groups) from doing stupid things.

Biases That Cause Bad Decisions

Biases come in all shapes and sizes. Two common types can sabotage your good decision-making efforts:

- **Cognitive biases.** Your brain regularly takes shortcuts to make sense of the world. These shortcuts help humans process tons of information quickly. This can be hugely beneficial. However, when these shortcuts go wrong, they lead to errors in thinking or judgment.

- **Data biases.** When data isn't a fair or accurate representation of the whole picture, it can lead to wrong conclusions or bad decisions. Data can be collected or analyzed in a biased way.

The Brain's Sneaky Shortcuts

Some predictable cognitive biases that can prevent you from making top-quality decisions are : confirmation bias and "cherry picking," the Texas Sharpshooter fallacy and the Clustering Illusion, data dredging and commitment bias, hindsight bias and overconfidence, and cognitive dissonance and Goodhart's Law.

CONFIRMATION BIAS AND "CHERRY PICKING"

Confirmation bias happens when people give more weight to information that supports something they already believe. When presented with information that goes against their belief, they downplay or overlook it. "Cherry picking" is similar, but it's done with more awareness. It's when people only show data that proves their hypothesis and hide data that goes against it.

To overcome both confirmation bias and cherry picking, remember your position on the null hypothesis. You must see very convincing evidence to sway you from your default belief. You can also ask someone to play a "devil's advocate" to argue the opposite side of a belief—even if they don't really believe it—simply to stress-test an idea. When you bring other people into the conversion and think with you, it ensures that your conclusion is well founded, the weaknesses are identified, and all angles are considered. To avoid cherry picking, show all the data and tell all the stories that could explain the experiment's outcome and follow the one that has the most supporting evidence.

TEXAS SHARPSHOOTER FALLACY AND THE CLUSTERING ILLUSION

The Texas Sharpshooter fallacy is when someone changes their goal or plan *after* seeing the results. Changing your goal like this is akin to a Texan firing a gun wildly at the side of a barn. Then, to prove they're an accurate marksman, the "sharpshooter" draws a target around a random cluster of bullet holes. Good sharpshooters set their target *before* firing—not after as shown in Figure 8.1.

FIGURE 8.1 Don't choose or change your target metric or goal after you see the results.

This is related to the Clustering Illusion. This illusion points out that in random sets of data, clusters will always form. This is illustrated in Figure 8.2, which contains a random set of data points.

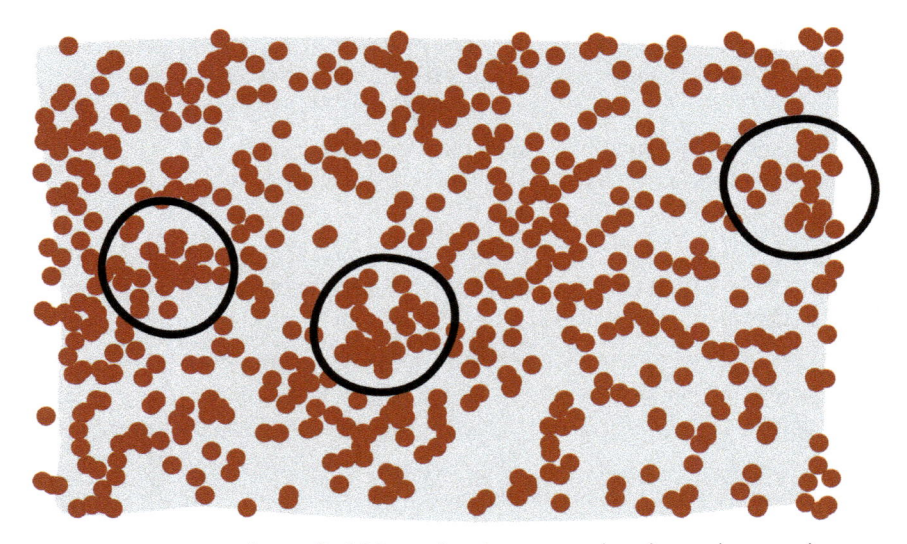

FIGURE 8.2 Human brains find "clusters" in data, even when the results are truly random.

In A/B testing, if someone engages in Texas Sharpshooting, it could mean that they change their test type, success metric, or hypothesis *after* seeing the results. A common mistake is aiming for a measurable improvement by running a superiority test. However, when the result shows no measurable effect, people keep the change anyway. In effect, that changes the superiority test to a non-inferiority test. They've changed the aim of the test *after* seeing the result.

To avoid this, design your experiment plan before you run your test and stick to it. The Validate template in the Appendix accounts for people's tendencies to bend information to their will by documenting important decisions. Because you fill out the template *before* you start your test, you'll notice if your brain tries to fudge a different result. But don't always ignore interesting yet unexpected results. Although it's likely due to randomness or your brain's tendency to find patterns, it could also be a signal of something bigger going on.

> **TIP** LEARN WHEN YOU GET NEW INFORMATION
>
> If you learn something new from your experiment, it's good to change your mind! But it's important to understand the reliability of that information and the trade-offs of making decisions with it. To stay safe yet not overlook a potentially strong signal, use an interesting result you see in an experiment to form a new hypothesis. Then run a fresh experiment. If the result is real, you'll likely see the same (or similar) result. Now you can roll it out with confidence. Or you'll learn if it was a fluke. ∎

DATA DREDGING AND COMMITMENT BIAS

Data dredging (aka *p-hacking*) is when someone tweaks their data or process to get a result they want—even if it's not valid. In experimentation, it means you create and search through large pools of data to look for interesting patterns without a plan. Imagine you have an experiment that shows an inconclusive result on your main metric. If you were to data dredge, you might search for different segments within your sample that show the result you aimed for. Then you would change your hypothesis and goal to align with the data you have. Data dredging increases the likelihood of observing a false positive since data behaves in a volatile way. Examples of data dredging behaviors that can lead to bad decisions are:

- **Running many tests on the same concept.** At some point, if you run enough tests, you will observe a false positive—when a test suggests there's an effect when there really isn't.

- **Running tests with many variants.** Similar to running many different tests on the same idea, this behavior also increases your chance of observing a false positive.

- **Spaghetti testing.** This comes from a funny saying about "throwing spaghetti against the wall to see what sticks." In spaghetti testing, a person throws many tests against the wall to see "what sticks." Spaghetti tests usually have poor hypotheses, aren't typically based on user research, and focus on "moving fast." This behavior leads to the *perception* of progress—without really moving forward—since false positives will inevitably happen.

Engaging in data dredging is easy when you're really invested in a test idea. Perhaps a test took a lot of time to make, the research was expensive, or a senior manager loved it. Running and re-running a test because a lot's been invested into it—even if that thing proves to be useless or bad—is a form of commitment bias. To avoid this trap, follow this simple process:

- Run simple A/B tests with a low significance level (alpha).[2] Make sure that you only test a thoughtful execution that you believe in.
- If an idea fails, don't iterate endlessly. Pick one or two of your best follow-up ideas and make them bug- and usability-issue-free.
- Stop pursuing the idea if you don't get a signal that you're on the right path.

HINDSIGHT BIAS AND OVERCONFIDENCE

Hindsight bias is when people prance around after they see a result with an "I-knew-it-all-along" attitude. After you know the outcome of something, your confidence in it (and yourself) skyrockets. The trouble with this attitude is that a result could be positive for many different reasons—not only the reason associated with your hypothesis. Just because you observe an effect that's likely real, and just because you have a logical hypothesis that offers a reason as to why—the effect could *still* be caused by something else. This overconfidence can seep its way into other experiment analyses causing you to be less rigorous in your analysis. Don't get overconfident when you get a few wins. Keep an open mind to different interpretations of the data. And *always* consider all the possibilities about *why* something happened.

When the results are negative, you double-check everything. When the results are positive, many people assume they must be right.

---

2. If this hurts your brain, it's because how significance levels (alpha) work is counterintuitive. Lower significance level is better because it makes the threshold harder to hit, which increases your confidence. This is almost as confusing as the confusion matrix!

COGNITIVE DISSONANCE AND GOODHART'S LAW

To understand other ways your brain might trick you, look at your work environment. Company culture and incentives dramatically shape behavior. If people's compensation is tied to their perceived "success" rate, they can do mental gymnastics to justify questionable decisions. Perhaps there's an experiment that's good for the business short term, but qualitative feedback can indicate it damages brand perception. The tension created between the incentive and the possible interpretations create discomfort. That discomfort, when ethics don't align with incentives, is called *cognitive dissonance*.

Another problem related to cognitive dissonance is when a company laser-focuses on a single or select few metrics to the detriment of all other ones. This imbalance causes a form of intellectual laziness called *Goodhart's Law*.[3] Goodhart's Law states that "when a measure becomes a target, it ceases to be a good measure." If a company obsesses over one measurement, it might miss the bigger picture. Setting a singular target and stripping the thought from the task encourages behavior that doesn't truly benefit the business. People will find ways to seemingly boost that number without actually improving the overall goal and having the intended impact.

One way to overcome these problems is not to attach compensation and other recognition to success rate—especially on a single metric. Trust employees to embrace and navigate the complexity that comes with making impactful business decisions. Then design a reward system that reinforces trustworthy and detail-oriented behavior.

Connecting the Wrong Dots

Just as your own brain leads you astray, data can, too. Whether it's because the data was collected in a biased way, or it's been presented in a biased way, making decisions with biased data is a bad idea. To make good decisions, collect, consider, and connect all the dots in a healthy dataset. Beware of the following data biases that can keep you from getting the full picture of a change's impact.

---

[3]. Paul Evans, "Money, Output and Goodhart's Law: The U.S. Experience," *The Review of Economics and Statistics*, 67, no. 1, (February 1985): 1–8, www.jstor.org/stable/1928428

A FUZZY "WHY"

Scarcity can nudge people to decide faster than normal. It's also social proof, because people think, "If it's almost gone, that means other people chose it, too!" In one experiment example, a team added stock availability to a search results page.

They used a few simple words, which said, "Only 1 left!" or "Only 5 available!" The number of items shown depended on how many items were available. If stock was scarce, the information nudged people to decide sooner rather than later. That nudge caused a sales increase.

Later, during a usability study, a participant asked, "Does this mean there's only *one* of these left *everywhere*? Like once it's gone, it's gone?" The designer's mind was blown. To them, it was clear that the stock level was specific to the website. That's because the designer understood the system. They knew where the number was pulled from and how it got there. The participant, however, didn't have that context, which left room for interpretation.

After the designer learned the copy wasn't clear, they ran a test. The line now said, "Only 1 left on our site!" or "Only 5 left on our site!" Again, the number depended on the stock level in the system. The test looked something like Figure 8.3.

The result of the change was yet another increase in sales. But debate about why sales increased sparked immediately. The designer believed the "why" behind the change was because it added clarity—because that's what they hypothesized. Other designers suspected that the longer text in its bright color made the persuasive message even more noticeable. Finally, another group thought an improved interaction (due to the longer text) increased people reading the tooltip, which explained how the count worked. Although the hypothesis was simple and the results were clear, the why was still fuzzy because of factors that couldn't be pulled apart. That's why—even if you think you know why something happened—you should stay open to other possibilities. Things aren't always what they seem, and that's part of what makes experimenting fun!

FIGURE 8.3 The control (top) said "only x left!" The treatment (bottom) said "only x left on our site!"

THE DECIDE PHASE: MAKE AN OPTIMAL CHOICE · 211

THE NOVELTY EFFECT AND THE CHANGE CURVE

A novelty effect (also known as a *burn-in effect*) usually happens when a large portion of your users are returning visitors. It might mean that users interact more with a new feature initially because they're curious and new things release dopamine in human brains—not necessarily because it's better. It usually happens when you add something new vs. change something that already exists. This effect is usually temporary, and it wears off in a few days to a few months.

The Kübler-Ross Change Curve causes a data bias similar to the novelty effect, but for a different reason: people don't like change.[4] The model in Figure 8.4 captures the idea developed by psychologist and author, Dr. Elisabeth Kübler-Ross and popularized in her 1969 book *On Death and Dying*.

THE CHANGE CURVE

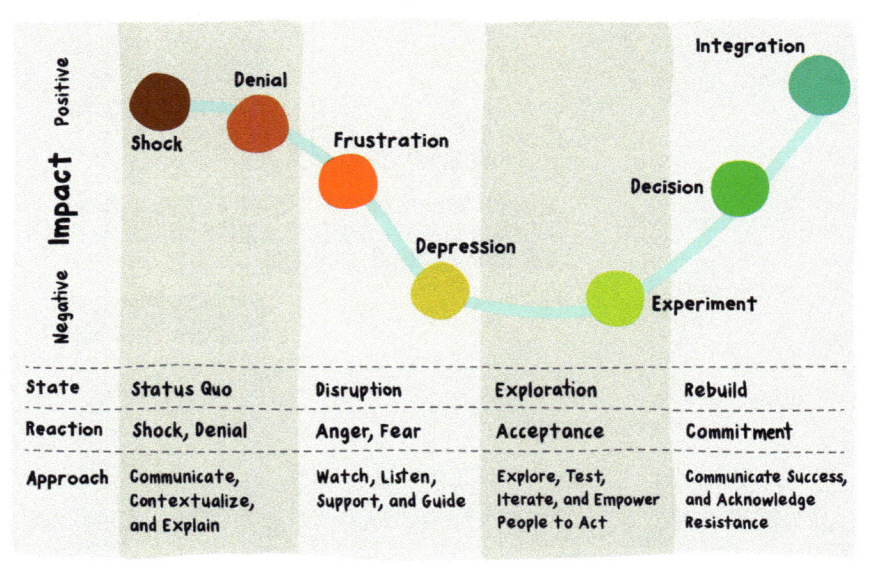

State	Status Quo	Disruption	Exploration	Rebuild
Reaction	Shock, Denial	Anger, Fear	Acceptance	Commitment
Approach	Communicate, Contextualize, and Explain	Watch, Listen, Support, and Guide	Explore, Test, Iterate, and Empower People to Act	Communicate Success, and Acknowledge Resistance

FIGURE 8.4 The Change Curve has four stages: Status Quo, Disruption, Exploration, and Rebuild.

---

4. The Mindtools Content Team, "The Change Curve," Mindtools, www.mindtools.com/au03rgg/the-change-curve

The Change Curve captures the stages that people go through when they adapt to change. People may experience the Change Curve when they must learn a new interaction. Until that learning happens, it might seem that an improvement was actually a step back. This effect especially impacts regular users of complex applications. At a certain point, even if a task is complicated, it's remembered through "muscle memory" (i.e., *procedural memory*). It becomes almost automatic for people.

To get people through the Change Curve quicker, design your experiments to be easy and pleasant to use. Don't stray too far from their current mental models and give them help. And most importantly—do *whatever you can* to make sure that the "change" is an "improvement." To limit the amount of bias in your data, you can run experiments a bit longer. This gives people time to play around with and adjust to the new thing or design. The time you wait depends on how often people use your product. If people go a long time between uses, you might need to wait a long time for the effect to wear off. Conversely, if people use the product very often—the effect might wear off quickly. If you don't have much time, you can also see if different user groups (new vs. returning visitors) behave differently when exposed to the change. Novelty and change don't impact new users in the same way as regular users.

SELECTION BIAS

Selection bias is when experiment visitors aren't chosen randomly. Because of this specific selection, the results aren't broadly representative and can be skewed. One form of selection bias is to not balance weekday users with weekend users. It's not uncommon for users to behave differently during different days of the week. The devices they use, how often they use the thing, and their time spent might vary because of their schedules and environment. It could also be impacted by big events, such as holidays or natural disasters.

To overcome selection bias, make sure that you randomize properly. Don't select specific groups of people for your test (unless your hypothesis targets or impacts that specific user segment). Also make sure that you capture data in complete business cycles as discussed in Chapter 6. For some businesses, cycles are a full week, while for others they're a full month.

EXPERT ADVICE FROM THE FIELD

THE DANGERS OF "DIALING-UP" YOUR EXPERIMENTS WITH SEAN O'NEILL

I've been running experiments for decades across a wide range of e-commerce and advertising platforms. Many times, the tests are very straightforward and have obvious decisions at the end. But sometimes, the results are very confusing.

Back in 2015 at Tesco, we observed an experiment bias that could have led us to reject one of the highest performing improvements to the site. At the time, Tesco ran the world's largest grocery e-commerce business. Grocery home delivery is very different from the Amazon general merchandise e-commerce world I had come from. Rather than users shopping for one or two mid- to high-value and high-risk purchases, grocery customers purchased 30–50 low-price/low-risk items. Often, customers forgot some of their favorite essentials. Our goal was to reduce shopper friction to increase average basket size (which is a big factor in grocery delivery unit economics). We were testing out a product personalized recommendation widget right in the checkout pipeline flow called, "Have You Forgotten?" shown in Figure 8.5.

This served up a short list of personalized items the customer purchased before that were not already in their basket. For example, they usually bought oat milk, but they didn't have any in their basket. How about adding oat milk? The hypothesis seemed strong.

We ran the test for four weeks. The initial results were very positive! Shoppers who saw "Have You Forgotten?" (HYF?) added an average of two more items to their basket than those who didn't. That was a huge lift.

Upon deeper analysis, however, we were shocked by the erosion of customer lifetime value. People who used HYF? were shopping less frequently than people who didn't see it. That was very concerning because customer lifetime value is an important guardrail metric. So, we turned off HYF? because the financial risk was too big. We didn't understand why a feature we believed was useful led to people shopping less frequently with us.

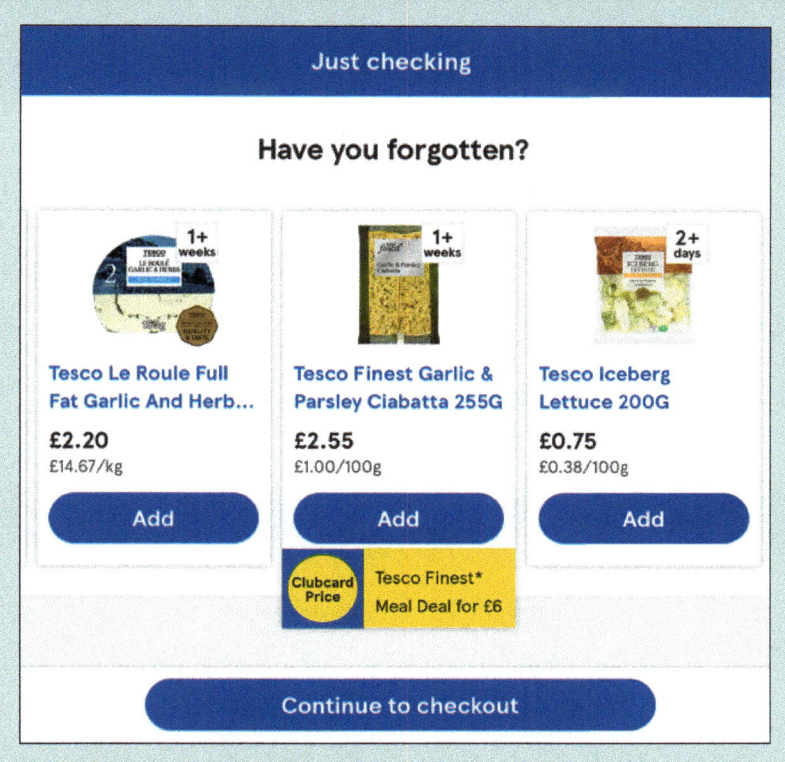

FIGURE 8.5 The experiment added a carousel of personalized items that customers usually purchased but that weren't in their baskets.

After months of deep analysis, we realized what caused the odd result. It turned out that the HYF? feature did statistically increase the basket size (even more than the initial test showed). However, the result that showed it reduced customer lifetime value was a false result. This false result was caused by a problem with the experiment design. The randomization was messed up, which led to a skewed allocation of users between control and treatment groups. We used an A/B testing tool to solve two distinct problems at the same time.

CONTINUES ➤

EXPERT ADVICE FROM THE FIELD

CONTINUED ➤

First, we used it to understand the financial impact of the new feature. This is what the A/B testing tool was designed to do. Second, we wanted to ensure that we didn't break anything (aka a "technical stability dial-up") as we released this incredibly complex new feature. To do this, we first released the feature to a small group of users—1%. Then, once we felt comfortable nothing was breaking, we very slowly and carefully dialed-up the sample ratio split to 50/50. This "dialing-up" from 1% to 50% over a period of time is how we unknowingly biased our data by capturing more loyal users in the control group than in the treatment. For grocery e-commerce, semi-loyal shoppers visit a few times a month. A super-loyal shopper visits multiple times a week. When the dial-up started with 99% of users in control and 1% in the treatment, we put 99% of our super-loyal customers into the control. Once a user is assigned to a treatment or control, they will never change their assignment for the duration of that experiment. As we increased the treatment session ratio over the next few weeks, 95% of our semi-loyal and super-loyal customers were already assigned to the control and stayed there. These loyal customers would never see the HYF? feature. We measured the results over the next four weeks. HYF? still performed dramatically well despite the HYF? treatment skewing mostly toward new-to-Tesco customers. The follow-up loyalty analysis of subsequent sessions per week and orders per week came out as negative for users exposed to HYF? because the super-loyal customers were all concentrated in the control.

This experiment design challenge of "user skew" affects most sites—especially ones with a lot of frequent users. To avoid biasing your data, always separate your technical stability dial-ups from your business experiments. Don't use the same experiment ID for both. Start and keep your sample ratio for a test at the level you designed for the whole experiment.

We were lucky in one respect: Despite the super concentration of super-loyal customers in the control, the HYF? performed so strongly it made us raise an eyebrow. This seemingly contradictory result that didn't jibe with our hypothesis was the only reason we dug into the results to learn more. The test could have easily been negative due to the heavy skew, and we might have thrown away one of the most impactful improvements because of an unintentional bias in our data.

Sean O'Neill
Chief Product and Technology Officer (CPTO)
at Syncron

Sean also served as the Chief Product Officer at GfK. He's run thousands of experiments in product leadership roles at large e-commerce companies such as Amazon and Tesco.

TWYMAN'S LAW

Tony Twyman was a psychologist and researcher who worked in the field of survey analysis. He worked for the BBC and was responsible for analyzing and interpreting large sets of data. Over the years, he noticed a pattern. Whenever a data point appeared to be particularly interesting or unusual, it often turned out to be wrong. Twyman's observation eventually became an informal rule among his coworkers, who started calling it Twyman's Law.[5, 6] Twyman's Law states that "any figure that looks interesting or different is usually wrong." Twyman's tongue-in-cheek observation highlights that you always need to look critically at your data—especially with surprising or unexpected results. When a number looks too good (or bad) to be true, take a second look. Chances are the result is due to a bug, a data collection issue, or a sample ratio mismatch. That's not to say that all interesting results are wrong. They simply need more rigorous and unbiased investigation to validate or invalidate the finding.

Infuse Some Objectivity

Objectivity is viewing something in a fair and unbiased way. Because experiments can feel personal, it's easy to get wrapped up in the excitement of finding "winning" tests. If that excitement (or behavior-shaping business incentives) influences decision-making, it will make your customer experience worse. Some techniques to remain objective and to focus on learning instead of winning are the following:

- **Practice cognitive distancing.** Cognitive distancing is when you mentally step back from your immediate reaction and look at a situation from an objective perspective. When you adopt the viewpoint of an external observer, you can get psychological distance from any mental or emotional attachment you have to an idea. One way to do this is to delay decision-making until your initial emotional response has dissipated. You should not have any emotional attachment to one

---

5. Ron Kohavi, Diane Tang, and Ya Xu, *Trustworthy Online Controlled Experiments: A Practical Guide to A/B Testing* (Cambridge, UK: Cambridge University Press, 2020), 39.
6. A. S. C. Ehrenberg and W. A. Twyman, "On Measuring Television Audiences," *Journal of the Royal Statistical Society*, Series A (General) 130, no. 1 (1967): 1–60, www.jstor.org/stable/2344037

decision over another when you make the final call. The ability to do this improves decision-making and lessens the chance for biases to win out.

- **Set up a review board.** Review boards are used in medicine, law, and academia. They are a group of experts in a subject matter who take an objective look at conclusions and supporting data practitioners provide. They provide guidance, feedback, and sometimes rulings on the outcome of whatever they review.
- **Do regular experiment reviews.** Take time to look at the design and results of experiments your team runs. Regular moments of reflection keep everyone focused on continuous improvement of the behaviors and processes to course-correct as needed.
- **Create a discussion forum.** A discussion forum can be a digital or physical space where people talk about decisions and their impact on the customer experience. Public embarrassment and shame are strong motivators. If people know there's a space where the quality of their work could be questioned, it nudges people to be a bit more rigorous in their thought process. This forum needs to consider all experiments and all experimenters. People might cherry-pick the experiments they choose to critique or share, so some curation and expectation setting is needed.

The risk of not setting up safeguards to shape good behavior is that people will (consciously or unconsciously) read the data in ways that benefit themselves. When this happens, it's often at the expense of the customer experience and long-term business success.

Systematic Reviews and Guardrail Metrics

Quantitative data doesn't give you the full picture. It can reliably tell you what and where something happens when you make a change, but it can't tell you why. Understanding the "why" behind a change requires a systematic review, and a healthy dose of good human judgment. A systematic review is when you combine results from multiple studies to get a clear overall picture of what's going on and why. This review helps you get more reliable and complete answers to your question. It's the highest form of evidence on the hierarchy of evidence (check it out in Chapter 1, Figure 1.3, in case you forget).

To do a systematic review, gather all the information you have about your topic of interest. Take the main findings from each study and see where they agree or disagree. By looking at all the evidence together and weighing it according to its reliability, you get a clearer overall picture of what's happening. When everything aligns, the decision is simple. You do what the data says is good for all stakeholders. When things don't align, making a good decision gets trickier. It's possible that your data shows a positive result on your targeted conversion metric but a negative result on your guardrail metric. It's also possible to get qualitative feedback from customers or an expert opinion that goes against your quantitative result. This is where a healthy dose of human judgment and more importantly—integrity—comes in.

Your job, when things are confusing, is to make an "optimal" decision. "Optimal," in this scenario, means to make the best decision possible given the circumstances. To do this, you need to balance all the pros and cons of each path forward and choose the one that brings the most benefit to your customers and (hopefully) the business as possible.

Weighing trade-offs to solve a problem for one while creating a problem for another is tricky business. All businesses are different, but the bigger and more complex they are, the more difficult it is to balance between stakeholder outcomes. For example, a food delivery company has four stakeholders to consider. They are the following:

- **Customer.** The person who orders the food.
- **Food partner.** The person or business that offers the food.
- **Delivery courier.** The person who takes the food to the customer.
- **Business.** The entity that sells the food and pairs the restaurant with a courier.

In this case, to make a positive business impact, you must carefully weigh the impact of the change on all the stakeholders and decide whose interest—if any—is more important than the other. What's good for one stakeholder, the customer, for example, might be harmful to another, such as the delivery couriers. A major challenge is to look for, work toward, and find the "win-win" solutions. "Win-wins" are decisions that equitably benefit all parties impacted by a decision. Either everyone benefits to the same degree, or a compromise balances the end result in a way that's suitable to everyone involved.

When Good Data Leads to Bad Design

When people put their own interests (or the interest of the business) before their customers, the consequence can be a badly designed product. Deceptive patterns and doom loops are two forms of manipulative UX patterns that arise regularly on the internet. Many people have a horror story or two about deceptive patterns or doom loops they've experienced.

Deceptive Patterns

Deceptive patterns are when users get tricked into doing something that's not in their best interest and usually at their expense. People design these patterns either consciously or unconsciously because the data shows a "positive" result on short-term conversion metrics. To avoid feeling "cognitive dissonance" that comes with doing harm to customers, people ignore or downplay the data that shows the harm. Familiarize yourself with different deceptive patterns so you can identify and call them out when they happen. Common deceptive patterns to avoid are:

- **False advertising.** This occurs when a product is described and advertised in a way that doesn't reflect reality. Perhaps it involves persuasive content and fancy photos to make something look better than it is. People discover this eventually, and there's often significant cost on the back end of implementing a bait-and-switch experience.

- **Forced continuity.** This is when, for example, a business requires credit-card information to start a "free" trial. When the trial ends, the card automatically gets charged without a reminder or an opportunity to cancel.

- **Roach motel.** Similar to forced continuity, this is when a business makes it very easy to do something, like sign up for a subscription, and almost impossible to get out of it. Roach motels get built because of poor or exploitative prioritization. That's why it's important to balance the short-term wins with the long-term gains that come with maintaining the trust of your customers.

- **Misdirection.** Businesses sometimes put *their* goal as the main path forward instead of the *customer's* goal. An example is making the customer's top goal a secondary call-to-action through the use of misleading color, size, and text. Then the primary action is what the business wants the customer to do.

- **Trick questions.** Questions or options can be presented in a way that makes users unintentionally choose an option they don't want. These trick questions often include unclear, confusing, or convoluted language, which makes people think more than necessary. For example, "Uncheck this box if you don't want to receive our newsletter."
- **Undesired defaults.** The power of defaults is real. But when you preselect a default option that's not beneficial to the customer (yet better for the business), it's a manipulative nudge. An example of this is preselecting the checkbox during a purchase flow to sign up for an email list when it's not the customer's goal. People are likely to overlook this kind of default, and it can even harm important metrics like your spam rate, which can mean fewer emails get through spam filters.
- **False urgency.** Creating a false sense of urgency is a pressure tactic that tips a design from persuasive to manipulative. Some stores artificially inflate their prices and then have a constant "sale" that never ends. This cheap tactic erodes trust with customers over time. It's also illegal in many countries and opens businesses to lawsuits and government intervention.
- **Disguised ads.** These happen when ads are made to look like other kinds of content—or even *navigation*. The result is that users mistakenly click on them, which artificially inflates click-through rates and might cost your company a lot of money on bad leads.
- **Confirmshaming.** This is when shaming language pushes users toward a choice. For example, a pop-up might offer a discount with the option, "Yes, I want the discount!" and the decline option as, "No, I like paying full price."
- **Hidden costs.** This happens when a customer is about to complete a purchase. Then, suddenly, additional fees or costs that weren't clearly disclosed earlier get tacked on to the price. At this point, the customer already went so far into the process that it's hard to back out, which is due to "escalation of commitment."
- **Sneaky add-ons.** This is when an extra product like a paid extended warranty, subscription service, or a related product is put into the basket without a customer's clear and direct consent.

- **Usability bugs.** When something is easy-to-use, it's typically lauded as a good thing. It's possible, however, to design things that are intentionally hard to use for the benefit of the business. Some people create usability bugs to encourage people to make money-generating or metric-improving mistakes. An example of this is by designing intentionally hard-to-close banner ads. As people try to hit the tiny "x" in the corner with an even smaller interaction target, they might accidentally tap into the ad instead.

Doom Loops

Doom loops are when customers get stuck in an unfortunate situation. They end up where they started without resolution or can't get past a certain point in a process. When customers get caught in a doom loop, they become frustrated and are likely to give up. Common doom loops that drive users to the brink of explosion are:

- **Phone service mazes.** This is when customers are forced to make a phone call to cancel a subscription or get help with a product. What's worse is that sometimes the caller has to pay a fee to be helped. Then they encounter an automated system that leads them through a series of options—none of which seem to apply to them. They end up being directed back to the website where they can't complete the task they were calling to complete. Or they get looped through the options again and again without ever reaching a human representative.

- **Endless FAQ redirects.** Sometimes when people need personalized assistance or an answer to a specific question, they get continually redirected to a general FAQ or help page. What they actually need is direct access to a customer service agent, which is continually denied by looping them through rabbit holes until they end up back where they started—on a FAQs page.

- **Claim submission barriers.** For services like insurance or product guarantees, customers might encounter a series of complex hurdles to clear when they try to submit a claim. They get led through many steps that require tons of documentation or repeated explanations. The amount of time, energy, and evidence to complete the task deters them from completing it.

- **Live chat limitations.** Companies love to offer "chat with customer service" options that end up being chatbots. When the chatbot doesn't facilitate a solution or answer the customer's question, it acts as a blocker to reaching a real human. It might even present a phone number to call, which then leads people through a dreaded phone service maze. Companies design these chatbot doom loops with the intent to reduce customer service costs under the guise of offering automated "help" or "self-service."

Do What's Right

To avoid deceptive patterns and doom loops, don't make trade-offs that benefit you or the business yet do harm to the customer. At the end of the day, you're responsible for the customer experience and the impact of your design changes. Choose wisely.

To help you and your colleagues make optimal decisions, lean on your experiment design and plan. The Decide template (in the Good Experimental Design toolkit in the Appendix) is specifically designed to shape ethical decision-making. It clearly defines up front all of the intended benefits (or the acceptable level of degradation) on different guardrail metrics for the different stakeholders. Have people complete and sign the Ethics section in the template before they put an experiment live. This reinforces the moral obligation that comes with designing things for people. Signing their name to the statement invokes a cognitive bias called the *commitment and consistency bias* to shape more ethical behavior. It's the opposite of *cognitive dissonance*, which is the feeling of discomfort people get when their actions don't align with their values or perception of themselves. When someone publicly commits to something, such as signing their name to a statement, they're more likely to follow through with their commitment to avoid the discomfort that comes with cognitive dissonance. People want to be (and appear to be) consistent in their actions and decisions. This desire becomes a force for good during the decision-making moment. This drive is *especially* true when the commitment is made voluntarily.

> **TIP** IS YOUR EXPERIMENT ETHICAL?
>
> Ask yourself these questions to determine if your experiment is ethical:
>
> - **"Would my mother be proud of me if she learned this was my experiment?"** You can replace "mother" with any ethical role model whose opinion you care about.[7]
> - **"Would you be embarrassed or proud to work here if this experiment succeeded?"** Ask this question to coworkers to check your quality of execution and impact on the company's brand reputation.[8]
> - **"What would the headline be if the design and results of this experiment hit the news tomorrow morning?"** This checks for legality and public ethical opinion. Imagine how the narrative could be skewed to examine all possible interpretations of the change. ∎

Dealing with Risky Business

A major benefit of Conversion Design is risk mitigation. If you find yourself making a high-risk business decision based on your data, it can make sense to revalidate your findings. Although the scientific process already drastically reduces your chance of making a bad decision, it's still possible to go wrong. Data, after all, doesn't give you the full story, and it can't *actually* predict the future flawlessly. It simply gives you more knowledge than you had previously to make a more informed decision. More ways to validate your findings are to:

- **Run the experiment again.** If the effect is real, it's highly likely that you'll see a similar result if you run the experiment again. Keep everything exactly as it was before to ensure a reliable result.

- **Replicate the concept elsewhere.** If the underlying hypothesis is the same, you can run a similar experiment with the same hypothesis but possibly a slightly different execution. For example, if you found a concept was highly effective on your website, it would make sense to try to replicate the concept in the same area on your mobile app if it's comparable enough. This is less reliable than rerunning the same experiment in the same area.

---

7. Hi, Mama. Thanks for holding a high ethical bar and keeping me honest.
8. My personal goal is to never embarrass my friend and colleague, Phil Hammel. He also holds a *very* high bar, but we all like him anyway.

- **Research similar findings.** Some concepts have been proven likely to work by being replicated across many different businesses. For example, it's commonly known that virtually anything that reduces page load time increases many types of conversion rates.[9]

When Good Ideas Fail

No one should test ideas they think are doomed to fail. Learn to craft the best variant possible to try and beat the baseline. But, alas—most things fail. Many large companies with experimentation programs have incredibly low "success" rates that typically span from 8% to 33%.[10] Smaller businesses might get a slightly higher rate, considering their products are likely less optimized. The point is—the odds are *not* in your favor. "Failing" in experimentation typically means two things:

- The change makes no measurable impact.
- The change makes a measurable, negative impact.

When a test fails, look for clues in your experiment data to understand if it's worth iterating, or if you should move on. Your options, depending on the situation, are the following:

- **Do more research.** Sometimes you see things in the data you don't totally understand. In these cases, ask your colleagues if they understand what's happening. They might have context you don't that explains it. Another option is to use the results to guide your follow-up research. What you learn might tell you where to look in your analytics, what you need to test more, or even if you need to start the Conversion Design cycle again. Experimentation can tell you what happens, but to understand the *why* even after a test, you might need to dig more.

- **Change something and try again.** Data sometimes gives clues about when to make an improvement and try again. If you see that an experiment does very well for many segments but not for others, learn why that might be. Once you get an understanding about what the issue is,

---

9. But remember what Jorden taught you in Chapter 5! At a certain point, you'll likely get diminishing returns on performance improvements.
10. Ron Kohavi, Alex Deng, and Lukas Vermeer, "A/B Testing Intuition Busters Common Misunderstandings in Online Controlled Experiments," 2022, https://bit.ly/ABTestingIntuitionBusters

it might inspire a way to improve the experiment, and you can run it again to see if the improvement pushes it over the edge.

- **Pass it on.** If you believe in the idea, but it didn't work, see if someone else wants to take a stab at it. A fresh set of eyes might provide the perspective and expertise to make it work based on what you learned so far. Don't pass things on without care and good reason, though. The more iterations something goes through, the more likely you are to see a false positive.

- **Drop it for now. Then try again later.** What didn't work yesterday might work today. Improvements in hardware, software, internet connections, and changes in user behavior might evolve enough to make your idea possible in the future.

- **Drop it and move on.** If there's no clear or good path forward, there's no harm in just trashing ideas and moving on to something else. Sometimes, winners need to quit to go win somewhere else.[11]

Positively Negative Results

When I see good ideas with conclusively negative results, I get almost as excited as if it were a win. This is because *most* things don't make an impact one way or another. When you find something that does anything at all, you learn that "something here *matters*." When you uncover something that matters—even by accidentally making it worse—that means you have the opportunity to make changes for the better.

After running tons of experiments I learned that you never really test a concept—**you test the *execution* of a concept.**[12] The success or failure of your idea comes down to how well it's implemented, and I've learned from experience that **there are far more ways to fail than there are to succeed.**[13]

---

11. Thanks to my dad for this advice. He's not a fan of the old adage, "If at first you don't succeed, try, try again." He has long advised me on the art of what we call "strategic quitting." We are inspired by the words of one of our favorite modern philosophers, Homer Simpson, who once said, "Trying is the first step toward failure."
12. I used bold here to emphasize how important this point is. So, please go read it again. This footnote is also so you can understand exactly how serious I am about this. Get it? Got it? Good.
13. This is also important enough to warrant the use of bold text. Seriously, get used to failure because there are tons of ways to accidentally make things worse. But don't worry—it's hard for everyone.

If a good idea fails, you could save it if you find a helpful signal in the data. Signals in the data can become the backbone for new hypotheses for you to make improvements and then try again. Over the years, I've seen these things make good ideas fail, get fixed, and then allow a good idea to succeed. Refer to the "When Good Ideas Fail" checklist in the Appendix to quickly triage the following common issues I look for first:

Poor Design

People untrained in design sometimes think it's easy. But it's not. It's possible that something as small as the spacing between elements or the wrong color choice impacts people enough to make a good idea fail. The role of visual and interaction design in experimentation is to make things accessible, easy to use, and to look good.

Edge-Case Bugs

Edge-case bugs are basically anything that doesn't work as it's intended to work because of a technical, visual, or interaction design reason. Track and measure as much as you can to identify broken things within your experiments. Lengthy and thorough quality assurance (QA) processes to ensure that products are bug-free are less and less common. When they do exist, live data feedback helps testers find and fix issues that slipped through the cracks. Things to watch out for are issues relating to specific browsers[14] and operating systems.

Technical Stuff[15]

People typically don't realize that an important part of a user's experience is often invisible to the eye. They spend so much time thinking about content, color, and placement that they ignore algorithms, errors, and speed. Bad code or shoddy implementation is another common reason that ideas fail. Technical things to look for when an idea fails are algorithm quality and its training data, page speed and the loading experience, and different types of errors and warnings.

---

14. I'm looking at you, Microsoft.
15. "Stuff" is a technical term I like to use.

Bad Localization

People from different cultures can have somewhat predictable preferences and common behaviors. This was discussed in Hofstede's Cultural Dimensions and the Culture Factor's Country Comparison Tool. Poor translations, values mismatched to cultural preferences, and broken interfaces can all make ideas fail. Look in your data to see if an experiment performs extremely well or poorly in any specific languages, regions, or currencies. This data can help you dissect if the issue is localization-related.

Poor Targeting

Not every experiment is useful to every kind of customer. People typically like to run experiments on 100% of their traffic, regardless of their context, to increase their sample size as much as possible. This makes sense most of the time. But if a change isn't relevant to all the visitors, your data gets "noisy." "Noise" in data is when a signal from a specific sender gets drowned out. But because that group is lumped in with everyone else, it's not possible to detect it.

Way to mitigate getting noise in your data is to use on-view, on-click, or on-scroll tracking, as well as back-end tracking. All of the "on" tracking ensures that people aren't tracked until they're actually *exposed* to the change, and it has the opportunity to influence their behavior. For example, imagine you have an experiment that is halfway down a screen but only 20% of people scroll that far down and are likely to see it. If you track people as soon as the page is loaded, that means 80% of people in your sample likely won't see the change you made. If you only track people who see the change you made "on-view," then you reduce that 80% noise in your data. Back-end tracking helps you measure any performance impact that could affect the result. For example, imagine that you have a new recommendation algorithm, but optimizing its performance would take a lot of technical investment. You can learn if investing time to optimize its performance is worth it or not by using both back-end and front-end (on-view, on-click, etc.) tracking. Back-end tracking captures the impact of any performance difference, and the front-end tracking captures the impact of the content that's shown.

If you're targeting a specific type of user, make sure that you are only tracking that type of user in your test. Common things to target tracking on are:

- Type of visitor (new, returning, post-purchase)
- Traffic source (direct, search results, email campaign, etc.)
- Authentication status (logged-in or logged-out)

Bad Timing

People in tech are typically way ahead of the average user when it comes to hardware upgrades and software updates. They tend to forget that not everyone looks forward to the next new thing. Back in the day when touchscreen devices became popular, many designers and engineers were quick to implement swipeable UI. The main place they used swiping was for image carousels. What seemed obvious to them was new and unnecessary to others. It's not just about people, though. Hardware, internet connections, and browsers might not be able to handle the new stuff reliably either. To make swipeable carousels work, typically a line of JavaScript or two is needed. Every new line of code opens up an opportunity to slow page speed, cause new errors, or any other number of issues.

Implementing a swipeable carousel seemed the obvious thing to do for my team who designed for touch devices. However, when we tested the swipeable carousel—they always failed. People weren't yet used to the swiping behavior, and the performance impact of the additional lines of JavaScript didn't make the interaction improvement worth the wait. It was only a few years later when most people interacted by swiping, and new implementations existed that reduced the performance impact, that the experiment succeeded. Seeing this happen over and over again led me to one of my biggest lessons learned from experimenting: Products must evolve at the same rate that people evolve. No faster, no slower. It's OK to test things again as devices improve, internet connections become stronger, and user behavior changes. The world evolves around your product, which can impact how people interact with it through the dimension of time.

Keep an Eye on the Impact

Once you've made your decision, keep an eye on the impact over time. Track your guardrail metrics—especially if they're a lagging indicator—to ensure that what you predicted will happen actually does. Eventually, as more and more people work in this way and make better and better decisions, you will see an impact in your company's performance. And don't assume that just because something was *good* at one point in time means that it's *good forever*. It's good to retest old concepts after time passes to see if they're still providing the value they had before.

To build a metric-savvy team, create and hang dashboards around so that everyone can quickly and easily see how different metrics are doing at any given moment. This not only reinforces the importance of improving them, but it also makes spotting issues with them faster. In addition to making important metrics prominent within a workspace, give people access to all the numbers they'd need to make a decision. Some publicly traded companies hide financial data and only show the percent increase or decrease. This can lead to bad decision-making if people can't calculate the absolute impact a decision might have on the business. If you trust your employees enough to make changes to your products, trust them enough to see all of the data those changes impact. Because at the end of the day, *you* as the conscious human capable of empathy, reason, and nuance make the decision—*not* the data. And you need to see the full picture to make the optimal decision.

The Important Bits

- Your job, after completing the entire Conversion Design cycle is to make an optimal decision. This means that you aim to maximize the benefit and minimize the harm to all of the stakeholders of your business.
- Many cognitive biases can impact your ability to make good decisions. Familiarize yourself with them and reflect on your motivations and thought processes before you come to a conclusion.
- You can never know without a doubt the reason why a test succeeds or fails. You can only have your own best guess based on the hypotheses you write. There is some joy in the wonder of never really being able to know something with 100% certainty. Embrace it and stay curious.
- Data biases can also creep in and sabotage your ability to make an optimal decision.
- After you complete the whole Conversion Design cycle, you likely have multiple forms of evidence that you can consider with different weights to inform your decision. This is called doing a *systematic review*. Don't just willy-nilly follow the numbers and ignore other valuable forms of evidence.

- Willy-nilly following the numbers and removing all form of human judgment in your decision-making process can lead to deceptive patterns and doom loops. Don't ship things that harm and annoy others simply because it makes or saves your business money. Remember: The goal is to make things *better*. Conversion Design, when you do it right, is actually pretty hard.
- Build a moment of ethical reflection into your experimentation process. Have the people running the experiment sign their name to the test so they feel a sense of ownership and responsibility for the impact it has.
- There are far more ways to fail at experimentation than there are to succeed. When your experiments fail, check to make sure that you haven't messed up your implementation because a concept is not the same thing as the execution of a concept.
- The world evolves around your product, which means what didn't work yesterday might work today and what worked yesterday might not work today. Time is an important dimension within experimentation, so challenge your past results as new hardware, software, and user behavior evolves.
- You, an actual human capable of empathy and reason, make the decision at the end of the day—not the data. Make the decision that would make a loved-one proud.

CONVERSION DESIGN PROCESS

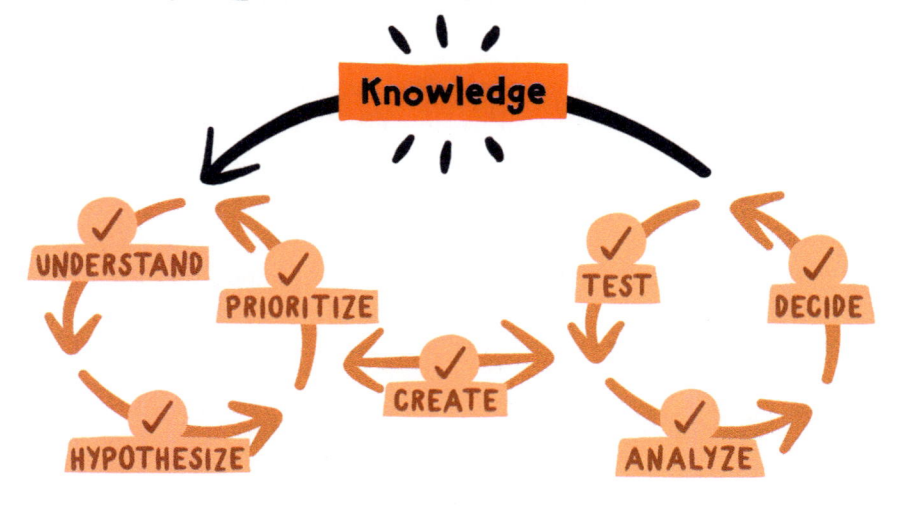

Knowledge

- UNDERSTAND
- PRIORITIZE
- HYPOTHESIZE
- CREATE
- TEST
- ANALYZE
- DECIDE

"Culture eats strategy for breakfast."
— Peter Drucker

CHAPTER

9

Scale It: Drive Impact Across Your Organization

THE POINT

Company culture—specifically an experimentation culture—is a powerful tool to scale employee behavior to maximize the business impact of Conversion Design. Design your company culture to reinforce the Conversion Design process.

Not long ago there was a company with an innovative and forward-thinking culture. It touted a hybrid work policy, unlimited vacation time, and cool swag for every new hire. At first, employees thought it was a cool place to work. But, after some months, they inevitably noticed that something was... *off*. The leaders always talked about the importance of "bold innovation." But, in reality, there was no out-of-the-box thinking. People kept ideas to themselves and did as they were told.

One day, a new employee excitedly pitched a bold idea to the boss. Instead of an enthusiastic response, the boss simply replied, "There's no room for that on the roadmap." Discouraged yet undeterred, the employee shared the idea with some coworkers. The coworkers excitedly joined in. Together, they took matters into their own hands. They made a quick-and-dirty implementation of the idea to A/B test, and it made a huge impact. They shared their first win with their boss, who was surprisingly pleased with the outcome. With leadership support, the idea was embraced and built upon. The brave heroes set the company on a new path to success.

Inspiring, huh? Too bad I made it up to illustrate my next point.[1] This narrative is how people *think* it will go when they pitch new ways of working. But reality bites. Experimentation advocates spend lots of time and energy pushing back against a well-oiled machine. Efforts to expand experimentation fail because employee behavior is shaped and reinforced by a company's culture. And culture is tricky. Boiled down, it's all of the beliefs, values, and attitudes internalized by employees and expressed through their behavior. Consistent behavior happens when it's reinforced. Behavior that's not reinforced slowly fades away. You shape behavior by positively reinforcing what you want and ignoring or negatively reinforcing what you don't want. To effectively shape human behavior to encourage consistent experimentation, you must learn to shape culture because it's a strong reinforcement mechanism.

The Four Aspects of Culture

Culture is the collectively accepted behavior and beliefs among a group of people. It consists of four things:

- Social norms and mores
- Traditions
- Artifacts
- Personalities

Social Norms and Mores

Norms are the basic behavioral expectations of a group. Obvious norms come in the form of explicit rules such as, "don't steal someone else's food

---

1. Womp womp.

from the work fridge." Sneakier expectations, however, reveal themselves over time. Most norms only become clear when someone breaks one. When someone does something that's considered outside of the social norm, it's called *social mores*. Mores are behaviors that are typically not acceptable or shunned within a group.

Traditions

Traditions are repeated group behaviors that define what the culture cares about. Celebrating traditions regularly builds collective habits, and collective habits shape outcomes. Or as Will Durant, an American philosopher, said, "We are what we repeatedly do. Excellence then, is not an act, but a habit." In this way, thoughtful company traditions pave the path to excellence. Not all traditions pave that path, though—most pave the path to mediocrity. Common examples of traditions are periodic celebrations, regular team activities, and promotion ceremonies.

Artifacts

Artifacts are both tangible and intangible aspects that set the tone of a culture. Tangible objects such as standing desks, beer in an office fridge, or bright brand colors set the tone of a physical space. Intangible artifacts are ideas such as a company's vision, values, and mission that set the tone of the working environment. These get introduced during new hire training, reiterated during company meetings, and referenced in day-to-day decision-making.

Personalities

Personalities are influential people within the culture. The behavior of these role models influences those around them. They're known to new hires and senior employees alike, and they often get sought out for advice or referenced in stories. Some of these personalities gain their influence through direct power, such as a CEO or director, while others gain it through indirect power, such as a charismatic individual contributor or a person who's considered a "culture carrier." To sum it up, all cultures can be identified and shaped by the four aspects in Table 9.1.

TABLE 9.1 FOUR ASPECTS OF CULTURE

NORMS & MORES	TRADITIONS	ARTIFACTS	PERSONALITIES
Social norms and mores are explicit guidance that rewards some behaviors and punishes others.	Repeated group activities, celebrations, or rituals.	Physical objects and written documents that illustrate, capture, and reinforce culture.	Prominent people who are looked up to as role models of cultural excellence and are sought out for guidance.
Examples: How meetings are conducted, who gets included in those meetings, and expectation of working hours and availability	Examples: Daily stand-ups, regular after-work events, coffee chats after lunch, end-of-year celebrations	Examples: Vision, mission, and values, posters and office decor, logo design, existence and type of company swag, tools	Examples: CEOs, employees who take on mythical status, onboarding leads, and other prominent figures

As Peter Drucker, a management consultant and author once said, "Culture eats strategy for breakfast." For Conversion Design to create a "Compound Effect," it must be acted on consistently at all levels of the organization over a long period of time. The common Drake meme in Figure 9.1 illustrates how strategy means nothing without a culture that reinforces it.

FIGURE 9.1 Strategy amplified by culture can have a big business impact.

Parts of an Experimentation Culture

To foster an experimentation culture, you need a foundation of trust and three key pillars:

- Equal access to information.
- Tools to enable action.
- Decentralized, democratic decision-making.

Healthy group dynamics are built on trust—and this is especially true for experimentation cultures. Employees in mature experimentation cultures get access to tons of data. Then they test their ideas with little-to-no oversight. From testing quickly, they learn quickly, and this fast learning is what makes a growth curve spike. In essence, in very mature experimentation cultures, employees are free and empowered to learn and know what they need to know so they can do what they think is best. Without these elements, a healthy and mature experimentation culture is unlikely. You can start building trust by opening up decision-making and seeing how it goes. Decision-making and trust go hand-in-hand, and so, they need to be addressed in parallel.

How to Build an Experimentation Culture

Building an experimentation culture is organizational change work. And change management is hard. Organizational culture is expressed through employees. So, for a culture to change—people need to change. Companies can do this in a couple of ways. One way is by influencing the people who work for them to change their behavior, and the other is by replacing those people with new staff. Usually, it's a combination of both. Culture-shaping exercises are best done together with employees to secure their buy-in up front. To set things off in the right direction, you must do the following:

- Hire good people.
- Trust the people you hire.
- Leverage cultural artifacts.
- Adopt new traditions.
- Shine the spotlight on the right personalities.
- Create new norms.
- Decentralize power.

Absolute Power Corrupts Absolutely

People often say that "knowledge is power." And that's totally true. But power is also many other things. When a company chooses to share access to information with its employees, it's decentralizing its power. When someone has access to information, they can turn it into knowledge (power). To achieve this, power dynamics must be carefully considered in your organizational design. Common types of organizations are hierarchical, matrix, and flat.

Flat structures support power dispersion better than a matrix or hierarchical ones. Hierarchical and matrix organizations are more resistant to change because power consolidates toward the top. People in positions of power rarely share it with others, unless there's a tangible benefit to them for doing so. In the words of historian and writer Lord Acton, "Absolute power corrupts absolutely."[2] It takes a strong and confident leader who believes in their people to willingly give up their power and disperse it among their employee base. That's why major change initiatives involving experimentation often require high-level leadership buy-in—especially in hierarchical and matrix organizations—because "change starts at the top."

However, even if you're not the CEO, you can still make waves if you understand power dynamics. To change power dynamics within a company, you need to understand the elements that define power. Once identified, you can assess how possible (or impossible) it is for you to influence them. A few elements of power to influence change are:

- **Information and knowledge.** This means different things in different contexts. In a business context, it means access to data and to people with different skill sets. Information gets turned into knowledge, and knowledge is where the real power lies. Gather as much knowledge as you can.

- **Inclusion.** This means access to who makes the decisions and how they get made. If you're not at the table where things get decided, you're unlikely to have the power to change how things are done. Do your best to get invited to those tables.

- **Money.** People and projects that get monetary support often get access to more resources and knowledge, which translates into opportunity. Try to get access to a budget to fund your change initiative.

---

2. "Lord Acton Quote Archive," Acton Institute, www.acton.org/research/lord-acton-quote-archive

- **Resources.** Resources is a broad category, and it represents anything from human energy and attention to physical space to tools. Get people on your side, take up space, and arrange access to the tools you need.
- **Perception and Reputation.** How people perceive people and ideas influences whether people will follow and align with an idea or personality—or dissent and look to distance themselves. Make sure that you frame your change initiative in a beneficial way to as many people as possible.

Employees in experimentation cultures are truly empowered to make decisions quickly and without approval. Not all leaders are willing to give up control, so leadership buy-in is integral to change.

Avoid "Game of Thrones"[3]

Some human character traits support experimentation cultures better than others. It's easier to hire people with these traits than it is to get people to change. As Jessica Herrin, founder of the successful jewelry company Stella & Dot says, "Shaping your culture is more than half done when you hire your team."

When you hire or expand your team, look for curious people with a blend of humility and confidence. Curious people who naturally want to understand things make great experimenters. As such, experimentation cultures aren't built with "yes men." They're built with "I wonder why…?" and "what if?" people. At first glance, humility and confidence might seem contradictory, but they are complements. Humility allows people to listen and accept when they're wrong, and confidence keeps them going when things get hard. It's possible you noticed some strong personalities in experimentation circles. These people stand steadfast behind their conclusions and state with conviction what they believe to be facts. However, these opinionated figureheads often have a fun quirk. When presented with evidence against their conclusion, they *think again*. This attitude is captured in the sentiment of "strong opinions, loosely held." The ability to change one's mind is a sign of intelligence, and as the Icelandic proverb goes, "A wise person changes their mind, a fool never will." That said, opinionated figureheads aren't necessary. They

---

3. I'm making a *Game of Thrones* reference here, but I've not actually seen more than the first five minutes of it. The only thing I remember before I quickly turned it off is incest and someone's head getting cut off.

are simply an exaggerated example of humility and confidence in action. To find curious people with the unique blend of humility, confidence, and product mindset, ask candidates these questions during an interview.

To test curiosity:

- **Have you learned anything new lately?** If so, what? Innovation happens when old ideas get combined in unique ways to create something new. The broader the knowledge someone has to draw from, the more unique ideas they can come up with.
- **How do you stay up-to-date with the progress in your craft?** This question checks for a dedication to continued learning.
- **If you could go back to school and study anything, what would you choose and why?** This indicates curiosity and helps you understand how broad their interests are.
- **What's your preferred learning style?** This question checks for people who have learned how to learn. It's not important for experimenters to know everything—they just need to know how to get the information they need when they need it.

To learn about humility:

- **What's the biggest mistake you've ever made?** What was the impact of that mistake? How did you manage the outcome? This helps you understand if the person regularly reflects on their behavior and learns from mistakes.
- **Talk to me about a time you experienced conflict at work.** What was the conflict about, and how did the situation unfold? This question checks for personal responsibility and a person's ability to view situations from different perspectives outside their own.
- **Show candidates various skill sets on a spectrum of competence ranging from novice to expert and then ask: Where would you rate your skill level with regard to this skill?** It's good to see people admit to varying degrees of skill across the spectrum. When someone rates themselves as expert or close to expert, this could be a red flag. Be sure to follow up to hear what evidence they have to support their claim. The ability to objectively assess their own skill and acknowledge where they have a lot to learn is a good indicator of humility.

- **Give an example of an ethically questionable experiment idea.** Ask them: What do you think about this experiment idea? This question assesses ethics, strength of character, and perspective-taking. Can they thoughtfully evaluate potential outcomes from various perspectives? Or do they default to right/wrong thinking? Can they navigate the gray areas and understand where different people's lines might be?

To gauge confidence:

- **What's something you're really good at?** How did you learn it? How long did it take until you were good at it? What you're looking for here is someone who acknowledges the hard work that's required to gain a skill. A red flag response is people who believe that skill is innate and don't acknowledge the time and practice involved with gaining expertise.

- **Talk to me about a successful project you're proud of.** How did you achieve that success? Could you do it again? This question checks for contextual awareness. Just because something worked for them in a specific setting doesn't mean it can be easily replicated elsewhere. Do they talk about what they learned and what knowledge might be applicable in another context? Or are they blindly confident that because they did it once, they can easily do it again in a different situation?

- **What's your greatest strength?** Does it ever tip into a weakness? If so, can you share an example of when a strength became a weakness? The world is filled with conundrums. Nothing is ever always good or always bad. There's usually a balance that's required to achieve optimum results. What you're looking for with this question is for people who speak candidly about how they maintain and sometimes lose control over that balance. And finally, if they have tools to find and keep the balance, that's a great sign.

- **What's something you changed your mind about recently?** What made you confident in that opinion before? And what got you to reconsider? This question shows self-awareness about one's own cognitive biases, as well as intellectual humility and flexibility. Do they think about how they think? Are they open to reassessing their beliefs? Human brains are tools, which need to be calibrated.

To understand commercial mindset:

- **Tell me about a time you discovered a customer problem.** This checks for curiosity about human-centered thinking and the capability to identify the ability to shift perspectives.
- **Did you manage to solve that problem? If so, how did you come up with the solution?** This illustrates the person's thought process and their ability to consistently find and solve real customer problems.
- **How did you know the problem was solved?** This assesses due diligence on solution validation and the person's ability to think about both customer experience and business impact.

Be especially careful when you assess confidence. Of course, no one wants to hire characters from *Game of Thrones*. But sometimes they sneak their way into the kingdom. It's better to shy away from candidates who are too confident than to take the risk. Curiosity, kindness, and enthusiasm can make up for what someone lacks in confidence.

As you speak with candidates, remember that interviews are a form of research, and job interviews are two-way research. Both the company and the candidate should use the time to learn about one another. Smart candidates don't miss this opportunity. Therefore, the questions the candidates ask you are also useful data to consider. Some useful things to look out for are: Did they put thought into what they wanted to learn from the conversation? Did they ask a common question that's already in the hiring process FAQs? Did they ask a question that was informed by research? You want people who take the time to think critically and distill what they hope to understand down into a few, well-formed questions. If they can ask thoughtful questions to glean useful information during a job interview, they can do the same in the job.

Trust the Trustworthy

The three elements in Figure 9.2 need to exist for trust to form. They are empathy, sharing logic, and authenticity.[4, 5]

---

4. Francis Frei, "How to Build (and Rebuild) Trust," TED, April 2018, www.ted.com/talks/frances_frei_how_to_build_and_rebuild_trust/transcript
5. Francis X. Frei and Anne Morriss, "Everything Starts with Trust," *Harvard Business Review*, May–June 2020, https://hbr.org/2020/05/begin-with-trust

Authenticity
Genuineness And Consistency
"I experience the real you."

Logic
Quality Thinking And Clear Communication
"I know you can do it—your reasoning and judgment are sound."

The TRUST Triangle

Empathy
Seek, See, And Listen
"You understand my perspective, and I believe you care about me and my success."

From "Begin With Trust," by Frances Frei and Anne Morriss
Harvard Business Review May–June 2020

FIGURE 9.2 The three elements of trust to build a healthy culture.

Empathy is the ability to understand and share the feelings of others, and it enables perspective-taking. When someone understands the impact of their decisions on others, everyone's well-being is likely to be considered. When people worry that coworkers don't have the collective's best interest in mind, trust crumbles. To safeguard against this, companies need a strong vision and an inspiring common purpose. The desire to meet the needs of others above the needs of oneself earns a steady assumption of good intent. And when people assume good intent, relationships are resilient, and people can make mistakes and learn without fear. This safe environment supports an appetite for risk that's needed to innovate. The Decide template in the Good Experimental Design toolkit (in the Appendix) reinforces the expectation of collective benefit, and it is empathy in action.

The second element of trust is sharing logic. For others to trust, they need to see consistent evidence of quality. The Good Experimental Design toolkit is an artifact that makes each coworker's critical thinking accessible to all. This kind of intellectual vulnerability also acts as a support system. Everyone has access to the thinking behind each experiment, so those who need support get identified and helped. Additionally, those who stray from ethical norms are held accountable.

Finally, trust needs authenticity to thrive. Authenticity is when attitudes consistently align with behavior. If people have any reason to believe someone's motives aren't true, skepticism takes hold and destroys trust.

Leverage Cultural Artifacts

Vision, mission, and values are key to how a company functions. They are the "why," "what," and "how" behind the work. A company vision is typically a short, inspiring description of the company's aspiration. The *mission* is what the company *does*, and the *values* are what *guides* the company's day-to-day decision-making. Shifts to company foundations can have a big impact. Make sure that you rework your vision, mission, and values to align with building an experimentation culture.

Artifacts that reinforce the vision, mission, and values are job descriptions, core competencies, and levels and expectations frameworks. Ideally, these documents are written through the lens of the company foundations to send a consistent message. Job descriptions and expectations tend to be underused levers to create change. They usually require a lot of work to update. In some cases, there might even be legal requirements for changes to be approved by employee representation groups. This is another great reason to bring employees along from the very beginning. As such, companies typically make documents once and don't evolve them much. This is good for stability, which creates a healthy work environment, but a change initiative is a great reason to revisit them with employee input.

Other useful artifacts can be more routine, for example, triggered emails for when an experiment gets started so that everyone knows a new test is live in the product. Another example is an auto-enrolled chat group or community where experiment info gets announced.

Adopt New Traditions

It's easy to confuse "winning" tests as the point of experimentation. However, the real point is to learn. That's why finding ways to celebrate knowledge—not numbers—is an important tradition. Regular learning sessions can shine light on interesting test results—be they positive, negative, or insignificant. Milestone celebrations are another good tradition. These are a fun way to encourage people to do more tests. For example, you might host an event after your team hits its first 50, 100, or 1,000 experiments. Or take a workday to do something fun together. But more doesn't always mean better. Anytime behavior is incentivized, people game the system. Quality expectations need to be measured or stated so that people don't pick up bad behaviors to get the reward.

Amplify and Spotlight Role Models

Top-level leadership has a huge impact on company culture, which cannot be emphasized enough. Without consistent, public support from leaders, change efforts are futile. Executives need to understand and believe in the benefits of this way of working to get people to follow. So, a certain amount of time needs to be dedicated to getting executives up-to-speed with the process, its practicalities, and the benefits.

Role models who consistently show the desired behaviors need to be recognized and visible. It's important that the traits that the company touts as important are the ones that are rewarded with attention, compensation, and increased responsibility. A good way to ensure role model influence is by having these people train new hires. Giving newbies a smart, friendly person to guide them through their first months is a powerful way to set people on the right trajectory. Some workplaces even make employer branding videos that feature these important company characters.

Language is another important part of culture. Personalities in experimentation cultures should speak more in terms of "we" than "me" to keep the focus on collective benefit and away from individuals. This is critical because behavior from influential people gets unconsciously emulated by others. Make sure that whoever gets the spotlight has an attitude worth amplifying. One way to know if a personality's message lands is through humor. Organizational "inside jokes" are a powerful way to communicate important messages and make people feel like part of the group. Habits form from consistent reminders and regular repetition. As such, repeat taglines and topics so often that people start to laugh.

Create New Norms

Norms are a very powerful culture-shaping tool. They're the regular behaviors that create a culture. Some useful behaviors to normalize are announcing experiments broadly, sharing analysis openly, and wrapping every product change in an experiment. This openness allows the culture to self-regulate for the desired outcomes. And wrapping every change in an experiment shows how even "harmless" changes can break things or impact customer behavior in weird ways. It also builds the experimentation muscle and codifies experimenting as the default way to ship changes and new code.

Another clever norm to adopt is using a "failure budget." Good experimentation platforms and metric design make it possible to know the financial impact of a test—good, bad, or otherwise. When people think a test is performing poorly, they might stop it early so they don't "lose the business money." Stopping tests early—unless there's clearly a bug—can be more wasteful than spending a little cash to learn. When tests end too soon, all of the time spent to design and develop the test gets thrown in the trash. To encourage running tests to the end, you should remind people it's part of their "learning budget." Report on how much of the budget gets used and assure people that it's valuable. If there are no negative tests, then people aren't trying. Or they're not pushing product boundaries to find the next mountain to climb. That said, use your data intuition that you build over time to identify when you're likely observing a real effect that's actually very damaging.

The Experimentation Cultural Maturity Spectrum and Its Milestones

The Experimentation Cultural Maturity spectrum with its five milestones is a nuanced way to track an organization's experimentation maturity. Maturity intertwines closely with team structure and workflow design. Each milestone marks the successful adoption of experimentation on a "micro" level. In any group of people, there are both micro and macro cultures. Microcultures are a kind of subculture to the macro culture. There are all kinds of ways that people can identify with different cultures even within the same group. For example, within a company there could be dozens of microcultures. Each one typically is set and shaped by the department lead. Usually, each craft discipline has its own set of values, beliefs, traditions, and artifacts.

How well these microculture expressions fit in with the macro culture varies. Sometimes, leaders intentionally design different microcultures to have conflicting values to create a "healthy" tension between the ways of working. Even individual teams have their own culture. Assessing culture periodically is a decent measurement of experimentation culture maturity.

EXPERT ADVICE FROM THE FIELD

THE EXPERIMENTER'S TROJAN HORSE WITH TON WESSELING

People often rely on the famous "five-minute install" testing tools, such as Google Optimize.[6] These tools can be put on most any website by pasting a few lines of JavaScript in just "five minutes." They are great to show experimentation as a proof-of-concept. Unfortunately, these tools can be uninstalled in five minutes, too.

I've found the most effective way to build and keep an experimentation culture is to embed it in the engineering tools and workflow. This means that all technical aspects of experimenting get owned by the tech department. The tooling becomes an integral part of the product's release cycle. Engineering teams are used for testing, monitoring, and rolling back data when something doesn't go as planned. So, it's a small shift in mindset to test, monitor, and roll back product changes in experiments if they're shown to be detrimental.

Getting experimentation tooling owned by tech is no small task, so it's important to get top-level leadership support. But once it's baked into the engineering pipeline and release process, it's there. Even if the CTO changes every few years, experimentation is just a part of how things are done because of how the product gets built. Then it's a hard behavior and quality check to remove.

Ton Wesseling
Founder of Online Dialogue

Ton's been helping businesses improve their customer experience and bottom lines through experimentation for more than 20 years.

6. RIP Google Optimize. This was a free tool that many people ran their first split-tests with. It recently was deprecated by Google, and it is no longer available.

Milestone One: Outsourced Execution

At milestone one, companies often outsource experimentation (as illustrated in Figure 9.3) so as not to distract internal teams, unless it proves useful.

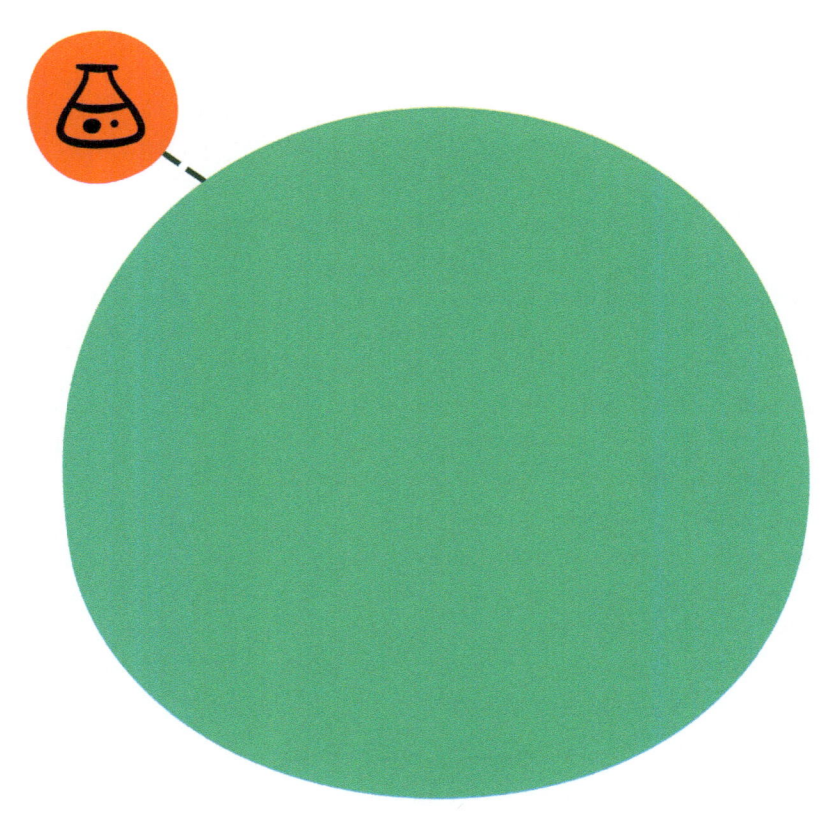

FIGURE 9.3 Milestone one of an experimentation culture is outsourcing the task to an agency or a consultant.

The number of experiments that get run depends on the contract and budget, and efforts typically start within the marketing department with a five-minute install testing tool. The contracting team relies heavily on the judgment of the contractor or agency, and the agency is incentivized to deliver results. Companies lean heavily into the concept of "quick wins" and copy what they find from other companies—with varying degrees of success. Experimentation efforts are usually siloed yet known and tracked by executives. If they decide experimentation is worth further investment, the company reaches milestone two.

Milestone Two: Internal Team

Milestone two is when at least one internal experimentation team (or a team that regularly runs experiments) exists. Some companies skip milestone one and jump straight to number two. This team may be supported by external agencies or consultants. They might also be tasked with ensuring that the data is reliable and getting the right tooling in place. They lean on simplified hypotheses and ship things when results are not only positive (superiority tests), but also insignificant (non-inferiority tests). This team tends to be siloed, and outsiders are generally unaware of their impact, as shown in Figure 9.4. Other teams might perceive this team as bypassing the usual release processes and delivery standards.

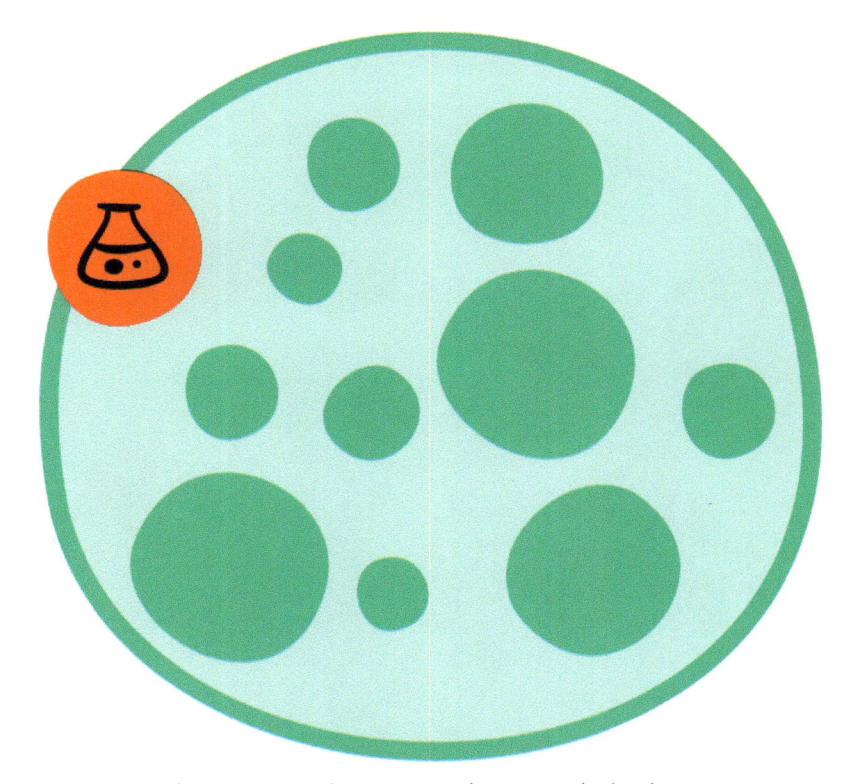

FIGURE 9.4 Milestone two is when an internal team is tasked with experimenting, often with support from a consultant or agency.

People on founding experimentation teams tend to be quite confident. They like to showcase their work, and they might think their method is superior. This attitude could cause some internal conflict, so keep an eye on inter-team dynamics. If the founding team is successful, they're often augmented with a technical team. Then the capability expands beyond the original team and influences other teams' ways of working. Unfortunately, because these efforts are pretty siloed, their positive impact quickly gets negated by teams outside of the area continuing with the old way of working. The Compound Effect is unlikely to take hold at this milestone.

Milestone Three: Business Function

Milestone three, illustrated in Figure 9.5, kicks in when a function expands the company's experimentation abilities beyond its own area.

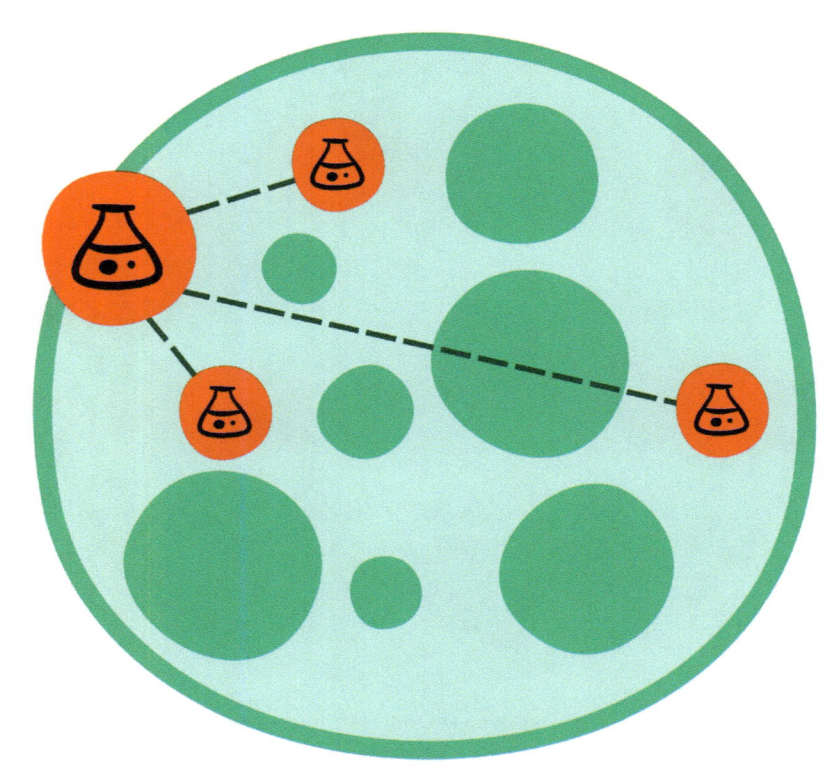

FIGURE 9.5 Milestone three is when an internal team starts to collaborate with others to socialize the new way of working.

The founding experimentation team runs their own tests and collaborates with others so they can learn, too. At this point, team focus should shift from skeptically leaning on quick-and-easy testing tools to searching for data quality and reliability. Teams at this stage often run A/A tests to check for situations where sample ratio mismatches (SRMs) are likely to happen. They find systemic fixes and minimize these issues so that teams can get reliable data.

Conversations about increased requirements and tool limitations arise. As such, the team may consider migrating from a simpler, client-side tool to a more robust back-end solution. More complex metrics are developed as teams develop important goals specific to their area of the product. Teams outside the central function rely on simple hypotheses and continue to ship experiments with a "do no measurable harm" attitude. Continuing to ship many non-inferiority tests, or worse—features that aren't A/B tested at all—blunts the potential benefit of harnessing the Compound Effect. The good decisions that get shipped are still negated heavily by things that are just released without strong quantitative evidence of benefit.

The founding experimentation team should understand the importance of the null hypothesis for data reliability and should shift focus to lay foundations to scale the practice.

Milestone Four: Federated Center of Excellence

When foundations are in place and scaling begins, a company is at milestone four, as shown in Figure 9.6.

At this point, teams can confidently set up experiments on their own. Ideally, people other than just engineers can set up and run tests to learn more and faster. The culture is supported with a federated structure. Federations have a center of excellence that is responsible for educating and advising on challenging decisions, although most of the time, teams have the freedom to govern themselves.

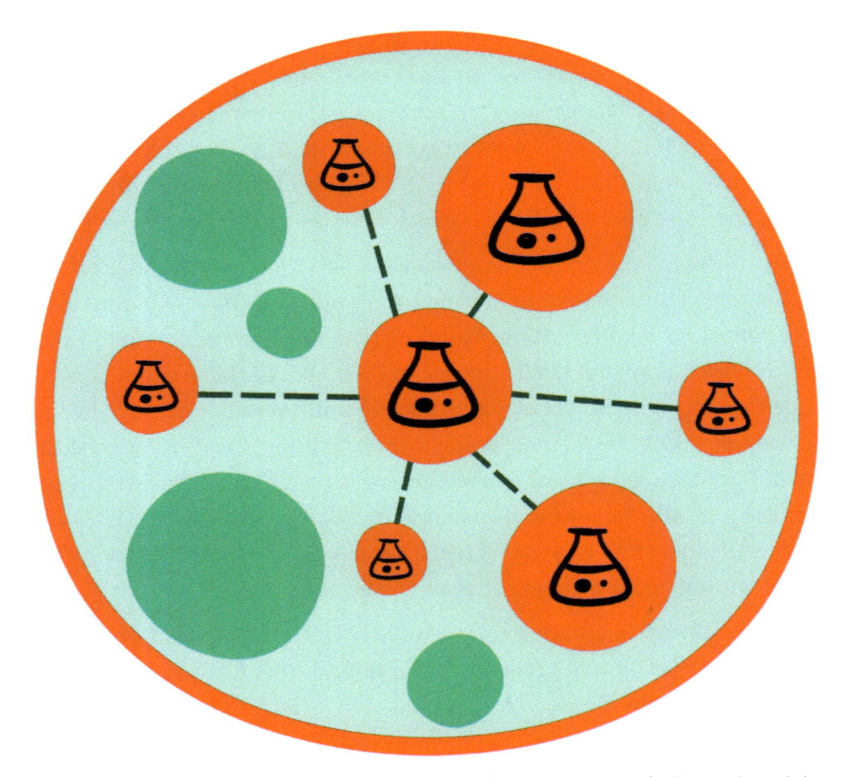

FIGURE 9.6 In milestone four, companies typically operate on a federated model.

Teams at milestone four are more likely to ship only positively impactful experiments, but they still ship non-inferiority changes. Experimentation tooling is reliable and robust. Metrics are varied and account for a mix of team goals. People trust results, and a team exists to build and improve an internal solution, or to customize an external one. CEO and executive input are used only for very challenging ethical, legal, and business decisions. Otherwise, teams make decisions as they see fit with the information they have. At this point, teams may start to see the Compound Effect starting to take hold on their business results.

Milestone Five: Decentralized Culture

A company reaches milestone five when all decisions get made with experiment data and only alternate hypotheses with a positive result get adopted. Tooling to track, measure, and run controlled experiments is available to everyone. Decision-making is democratized and mostly decentralized (as illustrated in Figure 9.7).

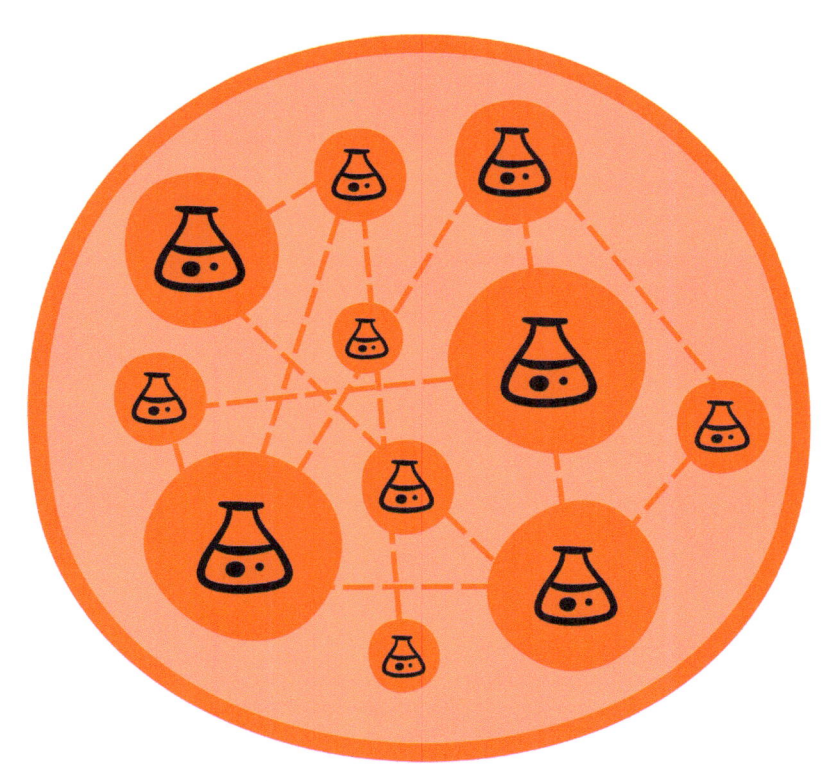

FIGURE 9.7 In milestone five, all teams are educated and empowered to experiment with little-to-no oversight from any governing body.

Hierarchy is flat and approval processes are minimal or nonexistent. Flat organizations only work with a strong, self-regulating culture supported by strong critical thinkers and craftspeople. Teams release all features, changes, bug-fixes, and campaigns in a controlled experiment. All employees have basic data literacy and need little support from a data scientist to analyze results. This approach frees up data scientists and machine-learning experts to find new ways to unlock other, more complex forms of experimentation. The Compound Effect is visible in the company results, and everyone contributes experiment ideas. Table 9.2 is an overview of all the experimentation culture milestones.[7]

TABLE 9.2 EXPERIMENTATION CULTURE MILESTONES

MILESTONE	MODEL	DECISION-MAKING
1	Outsourced	Decisions are made primarily by the outsourced agency while being influenced heavily by internal company stakeholders.
2	Internal Team	With support from an agency or internal expert who leads the effort.
3	Business Function	Primarily with the dedicated function as they support teams to learn the ability.
4	Federation	Balance between the team running the experiment and the function overseeing the efforts. Simple decisions are made at the team level, while more complex cases are brought to the "center of excellence."
5	Decentralized Culture	The flat structure supported by a strong experimentation culture allows for decentralized, democratic decision-making. Most decisions are made by the individual teams with little oversight. Only concepts that land squarely in the legally and ethically uncertain area get escalated to the central support function and executives.

7. This framework is not prescriptive. Some startups are founded with an experimentation mindset with all decisions tested and backed by evidence from day one. The larger the company, the more tests they can run without the culture being completely infused with an experimentation mindset.

EXPERT ADVICE FROM THE FIELD

CONNECTING CUSTOMER BEHAVIOR TO BUSINESS GOALS WITH DAVE MARTIN

Choosing your experiment goals is foundational to achieve the impact you aim for. It's typically very difficult to do this in a business-to-business (B2B) environment. Often, in these situations, the customer and business value are assumed. Assumptions are commonly held beliefs that don't need proof. Without assumptions, it would be impossible to progress. You'd be busy testing and retesting everything all the time. Because assumptions don't need proof, they sometimes cause problems, so you need to evaluate them continuously to ensure that they're working in the way you believe they are.

To pick the right business metrics to impact with experimentation, you need to craft clear value assumptions between user behavior and customer value. Then you tie the value assumptions between customer value and business value, which creates a Product Value Chain (Figure 9.8).[8]

FIGURE 9.8 The Product Value Chain gives teams customer behaviors to impact, which are assumed to have customer value, which translates into business value.

CONTINUES ➤

8. Andrea Saez and Dave Martin, *The Product Momentum Gap: Bringing Together Product Strategy and Customer Value* (RightToLeft Consulting, 2023), 39.

EXPERT ADVICE FROM THE FIELD

CONTINUED ➤

Product Value Chains tie a clear thread between user behavior through to business goals by way of customer value. Most product strategies fail to discuss user behavior. Instead, they typically consider the value the customer will enjoy and how that relates to the business goals. For example, a product strategy might describe a key initiative aimed to reduce the customer's operating costs, and in turn, expect increased payment for this. Hence, it supports a business goal of increasing revenue.

There's nothing wrong with this, but it's not always enough to support distributed decision-making. There are too many value assumptions left for people to determine for themselves. You need transparency on which user behaviors or activities the product can impact to reduce operating costs.

A brief example might be a product that handles invoice payments for small businesses. The value assumption could be that increasing speed of invoice reconciliation will reduce operating costs of chasing already paid invoices. The Product Value Chain in Figure 9.9 expresses what change in behavior is needed to create customer value, which then impacts the business goal.

Business Goals — **Customer Value** — **User Behavior**

Increased Revenue — **Reduced Operating Cost** — **Faster Invoice Reconciliation**

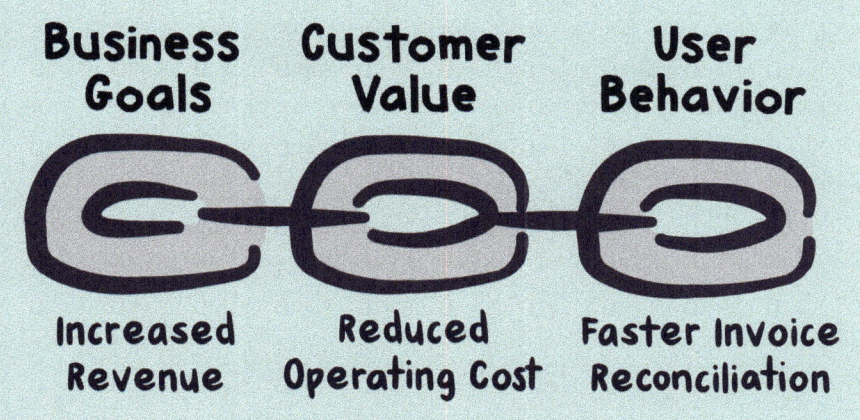

FIGURE 9.9 An example of an assumed Product Value Chain.

When product strategy is expressed in terms of user behavior to create or change, it's put into terms that product development teams can directly impact. But remember: Many Product Value Chains are based on value assumptions that also must be tested and validated over time.

Dave Martin
Founder of the product consultancy RightToLeft (www.RightToLeft.io)

Dave is also the co-author of the book, The Product Momentum Gap. *He specializes in helping business-to-business (B2B) companies improve their product development with a focus on strategy and leadership.*

Don't Be a Hero—Get Some Help

Organizational psychology is complex. People earn doctorate degrees to understand group dynamics and human motivation. So, it's smart to hire a trained professional. Change management specialists look at a company's current state to find the quickest path to success. If leadership agrees to adopt an experimentation culture, they'll be glad you asked for help. Leadership stamp-of-approval and a coordinated, professional effort can make all the difference.

One tool in an organizational psychologist's toolbox is an eight-step change management process. It was created by Dr. John Kotter, a Harvard Business School professor and author of 24 business books including, *Leading Change*. His eight-step framework illustrated in Figure 9.10 advises that changemakers do the following:[9]

1. **"Create a sense of urgency."** Humans tend to focus better with the pressure of a deadline. To foster a sense of determination, it's important to give a good reason and an attainable deadline.

2. **"Build a guiding coalition."** A common mistake that people make is that they try to change everyone all at once. This is a failing strategy. It's better to focus your energy on influential people who are likely to support the idea. Once they're on board, they can reach more people.

3. **"Form a strategic vision."** People need something to strive for, and a strategic vision provides this. When people know where they need to go and how to get there, it increases their chances of success.

4. **"Enlist a volunteer army."** Once a strategic vision exists, then it's go time. People like to be a part of something big, so now is when it makes sense to bring lots of people on board. Will you get everyone to join in? Nope. And that's fine. Again, focus your energy on people who are open to change.

---

9. John P. Kotter, PhD, *Leading Change* (Boston: Harvard Business Review Press, 2012), 23.

5. **"Enable action by removing barriers."** Having a plan and people are just two things needed to create change. It also needs to be possible. Remove any and all blockers to success so that teams can achieve their goals. You won't be able to see them all up front, but people need to know where to go for help when they come up against a blocker.
6. **"Generate short-term wins."** Success breeds more success. Earn and praise wins early on to give a feeling of progress and keep people engaged. Short-term milestones help to build momentum.
7. **"Sustain motivation."** Maintain momentum created by the short-term wins. Adjust perspective to aim for the mid- to longer-term goals of the change.
8. **"Institute change."** Ensure that people understand that success is tied to sustained behaviors. Adopt and reinforce important aspects of the new culture to sustain it for the long term.

1. Communicate Urgency
2. Build a Guiding Coalition
3. Form a Strategic Vision
4. Enlist a Volunteer Army
5. Enable Swift Action
6. Achieve Short-Term Goals
7. Sustain Motivation
8. Institute Change

FIGURE 9.10 John Kotter's eight-step change management process.

Address Complaints with Counter-Objections

"Experiments destroy the codebase" is a common complaint from engineers. For teams that do complex experiments, they usually need to add experiment tags and write extra code. And if there's one thing engineers hate—it's writing more code. To be fair, if experiments get run in a codebase like it's the Wild West, it can get pretty messy. But the solution to this problem is simple: *clean it up.*

For some reason, this isn't obvious. It might have something to do with the obsession of going "fast" and speedily moving from one thing to the next. Or, perhaps, people simply forget. Whether the mess is intentional or an accident, you can try these ways to make cleaning experiments easy and routine:

- **Isolate the experiment in one *code commit* or a separate branch.**[10] More experiments fail than succeed, so if engineers are smart about how they write their code, it's possible to find the commit or merged branch that introduced the experiment and revert it when it's done running. Cleaning up successful experiments requires a bit more thought, but they're also more gratifying to finish up.

- **Leave clear code comments.** Make sure to mark the beginning and end of any code that introduces an experiment or part of one. Label them with the experiment name so you can search for traces of it and easily know exactly what needs to be removed. Keep in mind, though, that code comments are visible in DevTools if someone inspects your code. Don't put anything in there that you wouldn't be comfortable with curious customers seeing.

- **Bake it into the process.** An experiment isn't done until it's been removed from the code. In this respect, it's useful to have a checklist of all tasks in the process that people can tick off as they work. Not only does this give people a sense of progress that keeps them motivated, but it also ensures that everything gets done in a healthy, clean way.

---

10. *Commit* and *branch* are engineering terms. If you don't know what I'm talking about, feel free to glaze over this section and just hand it to your developer.

- **Designate company cleaning days.** If baking experiment code cleanup into the development process is unsuccessful, setting aside time to clean everything all at once is another option. Depending on how many experiments your team runs, it could be once a quarter, once every half year, or once a year. If it's made into a fun, relaxing event that gets people away from the daily grind, it can also act as a culture-reinforcing company ritual.[11]

- **Set an expiration date on experiment tags.** Automated quality control is the greatest. It's possible to write a script that scours the code for old experiment tags. Every code base should have some kind of versioning system, which attributes lines of code to employees and dates. Once the script finds an experiment tag that was introduced longer than six months ago, it's probably safe to assume that it needs to be cleaned. After the script finds the experiment tags that are old and the employees who put them there, it can trigger an email or message bot to chase the engineer or engineer's manager (if you're not playin' around) until it's cleaned up. This is annoying yet effective.[12]

- **Write cleanliness into the job description.** Written expectations are powerful cultural artifacts. And job descriptions and career frameworks—both internally and externally facing—are two behavior-shaping tools. A performance assessment that's tied to leaving the codebase clean ensures a level of accountability for being messy.

Some of these ideas are "carrots" and others are "sticks."[13] With a well-balanced structure of fun, opportunity, incentive, and unfortunately—the sometimes-necessary deterrent of negative consequences—it's possible to keep code in good shape. So, if you ever hear an engineer push back against experiments because they're messy (or, if you are that engineer…), simply respond with, "Well, clean it, then."

---

11. Pro tip: Ordering pizza or getting cake for everyone on cleanup days gives people something to look forward to.
12. "Annoying yet effective" is how I describe myself at times, too.
13. I tried to think of a funny joke about "carrot sticks" and them being a confusing combination of reward and punishment, but I couldn't. Also, why are carrots considered a reward anyway? Cake seems like a better reward.

You'll likely come across many excuses as to why teams choose not to experiment. And they may or may not be legitimate. Some people don't like the "extra effort" involved, the time it takes to get the results, or the difficulty of actually achieving the impact the team aims for. Keep a running list of complaints people have and either address and fix them, or counter them with a different perspective. Changing people's beliefs and behaviors takes years of enthusiasm and effort. It's not a challenge for the weak of heart. But you can experiment and learn—so try different things in your situation to see what works!

The Important Bits

- Culture is shaped by and expressed through artifacts, norms, traditions, and personalities.
- Changing company culture is hard. It requires influence, which is a form of power.
- Power is many things. Some common attributes of power are information, knowledge, money, inclusion, resources, and reputation. Access to these makes culture change possible or impossible.
- Experimentation cultures are founded on trust and defined by an open access to information and decentralized, democratic decision-making. Leaders need to be supportive of this kind of culture for it to scale.
- It's easier to hire curious people who are humble yet confident than it is to get people to change.
- The Experimentation Cultural Maturity spectrum helps you know where your company is on its journey to being able to harness the power of the Compound Effect. The milestones are outsourced, internal team, business function, federation, and a decentralized culture.
- Use John Kotter's eight-step change management process to guide your change plan and hire a change manager to help.
- People will come up with all kinds of excuses not to run experiments. Prepare yourself by having a list of counter-objections to common ones to help keep excuses at bay.

CONVERSION DESIGN PROCESS

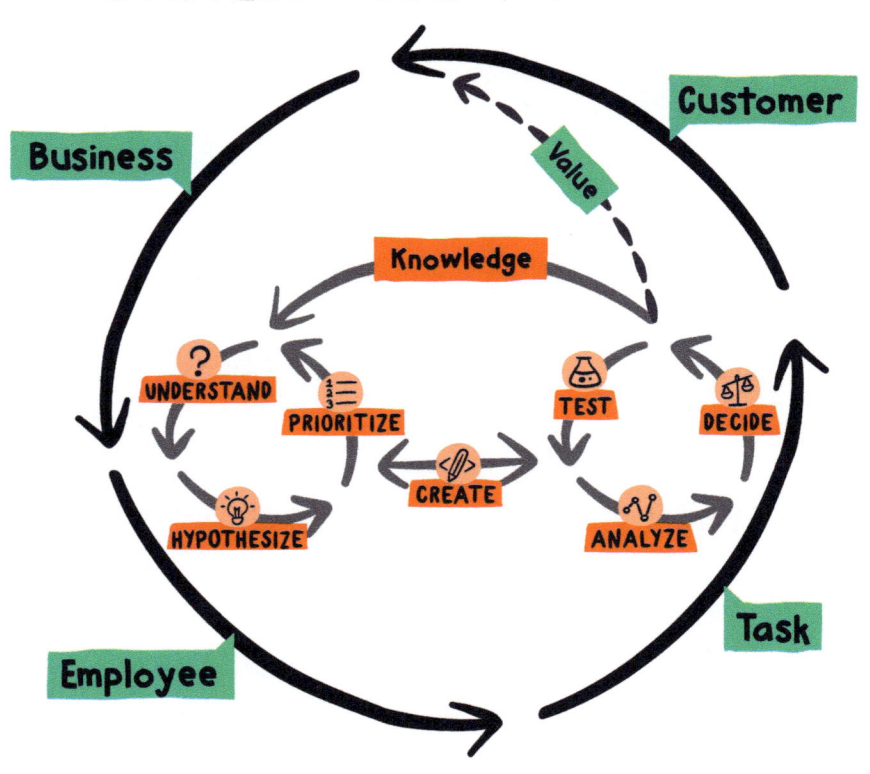

Final Thoughts

Hopefully, this book has given you a fresh perspective and some new knowledge to bring into your daily work. I encourage you to share the Conversion Design process with your teammates and bosses (or your employees if you're the boss!) to get other people on board. Because regardless of your job title—*everyone* has different strengths to bring to the Conversion Design table.

To that point, you likely noticed that this book covered a *lot* of topics: research, design, science, development, math, as well as business and product, in general. This was intentional. The goal was to give you the big picture behind Conversion Design and show you how to work across disciplines to design for impact *together*.

When you speak the language of your peers and understand the very basics of different disciplines, you become a more empathetic and inquisitive teammate. And connecting with people, asking questions, and learning stuff is at the heart of designing for impact. To truly leverage the Compound Effect, you and everyone in your business must test and learn as much as possible.

To set you up for success, I included the "Good Experimental Design" toolkit and a handy "When Good Ideas Fail" checklist in the Appendix. Bookmark these, use them, give me feedback on them (find me on LinkedIn or at www.erindoesthings.com), and share them. They are practical and living tools for you and your team.

And as a final thought, I leave you with this: *Anyone* can make things different. In fact, that's what most people do.

It takes courage and humility to question the value of your work so much to *literally* put it to the test. Testing your ideas is scary at first. And you'll likely feel all kinds of disappointment and confusion along the way. But in the words of one of my favorite human minds, Richard Feynman, **"There is pleasure in finding things out."**

So, go! Put your work to the test, find some things out, have fun, and make things *better*—not just different.

APPENDIX

The Good Experimental Design Toolkit

As Ronald Fisher learned (intro from Chapter 3, "The Hypothesize Phase"), experimental data is only as good as the design you put into it. This calls to mind a common saying among data scientists and software engineers: "Garbage in, garbage out." If the experiment has a poorly designed hypothesis—even if the test is randomized and controlled—it gives you garbage evidence. If your hypothesis is sound, but the math is bad, again—garbage data. To avoid creating garbage, follow the templates in the "The Good Experimental Design Toolkit." It contains four process templates, each with its own overarching theme to guide your approach.

Everything you need to know about how to fill out these templates is found within this book. These templates don't do the thinking for you. But they do structure your thinking in a way that makes it easier to design a good experiment. If at first you find filling the templates out hard or confusing—that's normal! The concepts they contain are difficult and require practice. Use this book as a reference until you get comfortable enough with the concepts to fill them out on your own.

Hypothesize: Design Like You're Right

"Design Like You're Right" reflects the confidence you must have in your experiment idea in order to test it in the first place. In the first template, shown in Figure A.1, you can define the following areas:

- **Baseline State:** Your default belief (the currently accepted theory) that you stubbornly stick to. You must observe convincing evidence to keep the change that's challenging the status quo.
- **Research Insight:** The observation that inspired the idea that will challenge the status quo.
- **Customer Problem Statement:** What you believe negatively impacts your customer as they try to achieve their goal.
- **Variant:** Your testable idea that will challenge the baseline state.
- **Prediction:** What you think will happen when the change is exposed to visitors. It's your proposed solution to the customer problem.
- **Variables:** The things that may influence the outcome of your experiment that you manipulate so you can determine causality.

HYPOTHESIZE

Design Like You're Right

Baseline State	The current state is... [description of base].
Research Insight	Based on... [research], [observation], or [evidence]...
Variant	We believe that... [description of testable idea].
Problem Statement	This is a [problem or opportunity] because... [assumption about the value it can create].
Prediction	If, we [proposed change] to [independent variable(s)], then [expected impact] on [dependent variable(s)].

FIGURE A.1 The Hypothesize template captures the thought process behind the change(s) you will test.

THE GOOD EXPERIMENTAL DESIGN TOOLKIT · 271

Validate: Test Like You're Wrong

"Test Like You're Wrong" flips the hypothesize hubris on its head by reiterating the extremely skeptical attitude you need when running a null hypothesis test. The second template, Validate, in Figure A.2, covers the following information:

- **The Null Hypothesis Reminder:** This reiterates your skeptical mindset, meaning you will not accept a new belief unless there's convincing evidence to prove otherwise.
- **Metrics & Math:** This outlines the exact evidence you'd need to see to be convinced to reject the null hypothesis.
- **Test Type:** This clarifies if you're aiming to make things better (superiority), or if your goal is simply to not make things worse (non-inferiority).

VALIDATE

Test Like You're Wrong

Null Hypothesis Reminder

The change is tested against the current state. The assumption is that the change has no effect.

However, if the effect outlined in Metrics & Math is observed, we will...

- change our minds,
- reject the current state, and
- adopt the change as the new current state.

Otherwise, we will reject the change and keep things in their current state.

Metrics & Math

This test is designed to find an impact on...

[_____] goal metric at a...
[____]% minimum detectable effect at a...
[____]% significance level and a...
[____]% statistical power after...
[_____] visitors, and...
[_____] run time.

Test Type

○ **Superiority**
(Make things better.)

○ **Non-inferiority**
(Don't make things worse.)

Decision Moment

Test will start on **[insert date]**.
Test will stop on **[insert date]**.

A decision will be made after the experiment runs for this pre-determined length of time.

FIGURE A.2 The Validate template template defines your decision-making criteria up-front to safeguard the reliability of the experiment results.

THE GOOD EXPERIMENTAL DESIGN TOOLKIT · 273

Create: Make with Care

"Make with Care" reminds you that you never purely test an idea—you always test the *execution* of an idea. Bugs and poor design decisions can doom your idea right out of the gate, so execution quality is key. The third template, Figure A.3, covers the following information:

- **Assumptions:** These are things you believe to be true that you have no evidence for. Making assumptions is a necessary part of learning because you cannot have evidence for every belief you hold.
- **Design Decisions:** Include any relevant information about the design decisions you made here. For example, explain why you chose a specific color that may deviate from a company color palette. The information you put here acts as a form of design documentation to help others learn about the execution of the idea you chose to test and why.
- **Development Decisions:** This block acts as your engineering documentation. Explain in this section what technology you used and why. For example, share what code language or tech stack you used.

The Create template is the foundation for your design and development documentation, which will help you during the Analyze phase. (Refer to Chapter 7.)

CREATE

Make With Care

Assumptions	In this experiment, we assumed… **[insert assumptions that guided the implementation]**.
Design Decisions	Because of these assumptions, we made these design decisions… **[insert relevant design decisions]**.
Development Decisions	To support the design, we made these engineering decisions… **[insert relevant engineering decisions]**.

Screenshots

Base | Variant

FIGURE A.3 The Create template makes clear what assumptions and implementation decisions were made in the test.

Decide: Do What's Right

"Do What's Right" forces you to consider the short- and long-term impact on all your stakeholders and the ethics behind your decision. The template in Figure A.4 reiterates the importance of finding the optimal outcome for everyone and everything impacted by the change. Consider the following things as you fill out the fourth template:

- **Stakeholder Benefits:** There's no such thing as one "right" decision in experimentation. It's your job to find the optimal path forward that benefits as many of your stakeholders as possible. This section defines your predictions about how the change will impact various stakeholders. This ensures that the net outcome of the experiment is as beneficial as possible.

- **Drivers & Guardrails:** These are the metrics you monitor to understand if you've changed customer behavior as intended and to ensure that no important business metrics are harmed.

- **Ethics:** This is a moment for teams to reflect on their motivations behind running the experiment. The act of signing one's name to an experiment is based on cognitive behavioral science. People want to behave in line with the image they have of themselves and willingly aligning with the ethics statement shapes behavior.

DECIDE

Do What's Right

Stakeholder Benefits

This will be good for **[stakeholder]** because...
This will be good for **[stakeholder]** because...
This will be good for **[stakeholder]** because...

Drivers & Guardrails

We anticipate a **[insert positive or negative change]** effect on **[driver metric]**.

We will not implement this change if it has **[insert % of positive or negative change(s)]** impact on **[insert guardrail metric(s)]**.

Ethics

I, **[your name]**, stand behind the ethics of this test, and I confirm that it is not

- misleading
- discriminatory
- manipulative

I understand that even if something is not illegal it can still be unethical. I am aware of biases that can influence my decisions, and I take ownership of the thinking and intent behind this experiment.

_____

Sign Here

FIGURE A.4 The Decide template to ensure that experiments are customer-centered, good for all stakeholders, and ethical.

CHECKLIST

When Good Ideas Fail

A non-exhaustive checklist to triage failed experiments.

DESIGN & EXPERIENCE

- ☐ Color contrast ratios
- ☐ Screen reader navigation
- ☐ Keyboard-only navigation
- ☐ Typography & legibility
- ☐ Color usage
- ☐ Information architecture
- ☐ Visual hierarchy
- ☐ Proximity & UI placement
- ☐ Content reading level
- ☐ Content scannability
- ☐ Translation accuracy
- ☐ Affordance & intuitiveness
- ☐ Framing & motivation
- ☐ Amount of content or choice

ENGINEERING & TECH

- ☐ Sample ratio mismatch (SRM)
- ☐ Page load time (web)
- ☐ Time to interaction (TTI for apps)
- ☐ Errors & warnings
- ☐ Animation performance
- ☐ Loading state
- ☐ Crash rate
- ☐ Layout shifts & flickers
- ☐ Algorithm quality & training data

AUDIENCE & CONTEXT

- ☐ Language
- ☐ Country
- ☐ Currency
- ☐ Payment methods
- ☐ Network speed
- ☐ User type (new vs. returning)
- ☐ Authentication status (signed in/out)
- ☐ Device type & screen size
- ☐ Browser type
- ☐ Operating system version
- ☐ Cultural preferences
- ☐ Purchasing motivation & objections
- ☐ Traffic source (direct, PPC, social)

Index

60/40 split, Impact Effort Matrix, 71–75

A

A/A testing, 76
A/B testing, 6–8, 54, 57
A/B Testing Tool, 143–145, 215–217
ABlyft, 190–192
ABsmartly, 190–192
absolute change calculations, 146–148
accessibility, 71, 92–96
Acton, Lord, 240
add-ons, sneaky, 222
advertising, deceptive patterns, 221–222
Aerts, Els, 32–34
aesthetics, 92–93, 113
Albrecht, Karl, 204
alpha, 160, 187–188
alt-text, 113
alternate hypothesis, 47–49, 50
analyze phase, 169–201
    bell curves, 180–183
    checking results, 191–192
    confidence intervals, 175, 183, 184–187
    conversion rate calculation, 175–176
    data volatility, 173–174
    foundations of analysis, 174–175
    interaction effects, 196–200
    looking and peeking, 171–173
    p-values, 175, 187–189, 190
    sampling distribution, 174–175, 176–179, 182–183
    statistical significance, 160, 187, 188–189
    as step in Conversion Design, 18
    visualizations, 189–195
Anderson, Kevin, 14–15
animation, 118
ANOVA (analysis of variance), 45
approval processes, 87
artifacts of company, 237, 238, 246
assumptions, 257
attitudinal insights, 25
authenticity, in trust triangle, 245, 246

B

back-end tracking, 59, 229–230
backlog tools, 84
bait-and-switch, 221
base/baseline, in testing, 6–7, 132
baseline conversion rate, 143–145, 149
because justification, 127
behavioral insights, 25
bell curves, 180–183
Bernardi, Lucas, 55–56, 139, 143
biases that cause bad decisions. *See* cognitive biases; data biases
bloat, as workflow problem, 83
blockers, as workflow problem, 81
Booking.com, 75, 120
Boone, Nekeia, 125
bottlenecks, as workflow problem, 80–81, 86
box plots, 194
break-even point, 162
breakdowns, as workflow problem, 82
bring your own device (BYOD), 27
Bristol, Muriel, 43–45
Bucher, Amy, 126
bug fix, non-inferiority, 169–171
bugs
    edge-case, 228
    as experiment ideas, 70–71
    usability, 223
burn-in effect, 212
business, as discipline of evidence, 3–4
business cycles, 162, 213
business goals, in Product Value Chain, 257–259
buttons and links, descriptive, 102

C

Cake Principle, 139
Catanuso, Francesca, 100–101
cause/effect relationship (causality), 53, 56, 137
Change Curve, 212–213
change management process, 260–261

chatbot doom loops, 224
cherry picking, 204–205
chunking, 95
claim submission barriers, 223
"click here" links, 102
click maps, 27–28
clicks, as goal, 62
Clustering Illusion, 206
codebase, cleaning experiments in, 262–264
cognitive biases, 126, 204–209
    cherry picking, 204–205
    Clustering Illusion, 206
    cognitive dissonance, 209, 221
    commitment bias, 208, 224
    confirmation bias, 204–205
    data dredging, 207–208
    Goodhart's Law, 209
    hindsight bias, 208
    overconfidence, 208
    Texas Sharpshooter fallacy, 205–206
cognitive dissonance, 209, 221
cognitive distancing, 218–219
collective intelligence, 17–18
collective knowledge, 9–10
collective stupidity, 204
commitment and consistency bias, 224
commitment bias, 208
communication, in workflow, 82, 87
company culture, 235–265
    aspects of culture, 236–238, 246–248
    cleaning experiments in the codebase, 262–264
    customer behavior and business goals, 257–259
    elements of trust, 244–246
    experimentation culture, 239
    hiring your team and interview questions, 241–244
    Maturity Spectrum, 248–256. *See also* Experimentation Cultural Maturity spectrum
    organizational psychology, 260–261
    power dynamics and types of organizations, 240–241
company vision, 246
Compound Effect, 16–17, 252, 253, 254, 256, 267
concepts, testing execution of, 227–228, 274
confidence, in team members, 241–244
confidence interval, 175, 183, 184–187
confidence level, 160, 175, 188–189
confirmation bias, 204–205
confirmshaming, 222

confounding variables, 53–54, 55, 56, 60
confusion matrix (error matrix), 150–158
content, as experiment ideas, 71
control, in testing, 6–7, 132
controlled variable, 54, 55
controlling confounds, 54
Conversion Design
    defined, 10–11
    design thinking and the scientific method, 46
Conversion Design, introduction, 1–19
    A/B testing in, 6–8
    and Compound Effect, 16–17
    evidence collection, 3–5
    and linear thinking, 8, 10, 14–15
    process, importance of, 17–18
    seven steps of, 18–19
    and systems thinking, 8–10
    value creation of, 10–13
Conversion Design process, 9
    and collective knowledge, 9–10, 20, 42, 66, 90, 130, 168, 202, 234
    with value cycle, 12–13, 266
conversion rate
    baseline, 143–145, 149
    calculation of, 175–176
    observed, 177–179
"conviction" of idea as high value, 72
country comparison tool, 99, 229
create phase, 91–129
    accessibility, usability, and aesthetics, 92–96, 113
    cultural differences, 98–99
    defaults and focus states, 103–105
    descriptive links and buttons, 102
    localization, 96–98, 229
    motivating people, 114–128. *See also* motivating people in create phase
    as step in Conversion Design, 18
    template: make with care, 274–275
    thinking for people, 105–113
cultural differences
    Hofstede's Cultural Dimensions and the Culture Factor's Country Comparison Tool, 98–99, 229
    localization, 96–98, 229
culture of company. *See* company culture
curiosity, 22–24, 241–242
currency format, and localization, 97
cursor color defaults, 104
customer confidence, 50, 52–53
customer lifetime value, 214–217

customer service contact logs, 29
customer service teams, and research, 35
customer top tasks, 23
customer value, in Product Value Chain, 257–259

D

dashboards, for teams, 231
data, exploring in research, 27–31
data biases, 204, 209, 212–218
    Change Curve, 212–213
    novelty effect, 212
    selection bias, 213
    technical stability dial-up, 214–217
    Twyman's Law, 218
data dredging, 207–208
data noise, 229
data points, in experience landscape, 36–41
data volatility, 173–174
dead ends, and error messages, 122–125
deceptive patterns, 221–223
decide phase, 203–233
    cognitive biases, 204–209
    data biases, 204, 209, 212–218
    deceptive patterns, 221–223
    doom loops, 223–224
    ethical decision-making, 224–225
    objectivity techniques, 218–220
    risk mitigation, 225–226
    as step in Conversion Design, 19
    template: do what's right, 276–277
    watching impact over time, 231
    when good ideas fail, 226–230, 278
decontextualized use of products in research, 25
defaults
    and focus states, 103–105
    undesired, 222
dependent variable, 53, 55, 137
descriptive links and buttons, 102
descriptive research questions, 24
design. *See also* Conversion Design
    as discipline of evidence, 3–4
    experimental, 43–45
    like you're right, hypothesize template, 270–271
The Design of Experiments (Fisher), 45
design thinking, in scientific method and Conversion Design, 46
detractors, users as, 29
devices in research, bring your own device (BYOD), 27

devil's advocate, 205
diminishing returns, 74, 120–121
discussion forums, 219
disguised ads, 222
dogfooding, 35
Don't Make Me Think (Krug), 105
doom loops, 223–224
Double Diamond design process, 14–15
double negatives, 153
drivers, 62–63, 276–277
Drucker, Peter, 234, 238
Durant, Will, 237

E

edge-case bugs, 228
effectiveness, in workflow, 77
efficiency, in workflow, 77–79
effort, impact vs., 71–75, 138–139
Einstein, Albert, 5
empathy, in trust triangle, 245
encouragement, motivating people, 128
Engaged: Designing for Behavior Change (Bucher), 126
engineering teams, and ownership of experiment code, 249, 262–263
entry surveys, 32–33
error. *See also* false negative; false positive
    in error matrix, 154–155
error matrix (confusion matrix), 150–158
error messages, 122–124
escalation of commitment, 222
ethical decision-making, 224–225, 276–277
    and cognitive dissonance, 209, 224
    deceptive patterns and doom loops, 221–224
evidence collection, 3–5
experience landscape, 36–41
experiment reviews, 219
experiment variables, 52–54
experimental design, and birth of modern statistics, 43–45
Experimentation Cultural Maturity spectrum, 248–256
    milestone 1: outsourced, 250
    milestone 2: internal team, 251–252
    milestone 3: business function, 252–253
    milestone 4: federation, 253–254
    milestone 5: decentralized culture, 255–256
    table of milestones, 256
experimentation culture, 239–248
    building it, 239
    company artifacts, 237, 238, 246
    elements of trust, 244–246

experimentation culture *(continued)*
   hiring your team and interview questions, 241–244
   norms, 247–248
   power dynamics and organization types, 240–241
   role models, 237–238, 247
   traditions, 237, 238, 246–247
experimenting, as research, 31
expert opinion, 4–5, 242
explanatory research questions, 24
exploratory research questions, 24, 31

F

failing, in experimentation, 226–230, 278
false advertising, 221
false negative (Type II error), 155, 160–161
false positive (Type I error), 155, 157, 160, 172–174, 207
false urgency, 222
FAQ redirects, 223
feedback, motivating people, 116–117
Feynman, Richard, 42, 49, 130, 139, 268
Figma, 75
Fisher, Ronald, 43–45, 47, 269
five-minute install testing tools, 249, 250
fixed horizon test, 174
flat-structured companies, 240, 256
focus, as quality of successful business, 86
focus states, and defaults, 103–105
Fogg, B. J., 93
Fogg Behavior Model, 93–94, 114
forced continuity, 221
form validation, 117
Four Bs (workflow problems), 79–83
   bloat, 83
   blockers, 81
   bottlenecks, 80–81
   breakdowns, 82
Four Buckets of Good Experiment Ideas, 69–71
framing techniques, 126
"free" label, 113
frequency, 177
front-end tracking, 59, 229–230
Fuller, John, 135
functional conflicts, 196–198, 200

G

Gauss, Carl Friedrich, 180
Gaussian curve, 180
Glaser, Milton, 90, 110

global maximum, 73–74
goal of experiment, as testing metric, 134, 135–137, 163
goals, in measuring success, 62–63
Good Experimental Design Toolkit, 269–278
   create, 274–275
   decide, 276–277
   hypothesize, 270–271
   validate, 272–273
   when good ideas fail, checklist, 278
Goodhart's Law, 209
Google Analytics, 27
Google Optimize, 249
Google PageSpeed Insights, 118–119
Google Trends, 100
grocery home delivery, 214
guardrails/guardrail metrics, 63, 120–121, 219–220, 231, 276–277
guessing like a scientist, 139

H

handoffs, 87
hardware upgrades, 230
Have You Forgotten? (HYF?), 214–217
heatmaps, 109
Hemingway App, 95–96
Herrin, Jessica, 241
Hess, Melina, 111
hidden costs, 222
hierarchical companies, 240
hierarchy of evidence, 4–5
hindsight bias, 208
Hofstede's Cultural Dimensions and the Culture Factor's Country Comparison Tool, 98–99, 229
Hopper, Grace, 5
horizon, fixing, 174
Horton, Sarah, 71
HTML tags, and defaults, 104
humility, in team members, 241–243
Hurston, Zora Neale, 20, 22
hypothesize phase, 43–64
   customer problems, 50–52
   experiment variables, 52–54
   measuring success, 62–63
   null and alternate hypotheses, 47–49, 50
   randomized controlled experiment, 54–55
   sample ratio mismatches (SRMs), 57–61
   scientific method, 45–46
   as step in Conversion Design, 18
   tea experiment, 43–45, 47
   template: design like you're right, 270–271

I

icons, for redundancy and navigation, 110–111
Impact Effort Matrix, 71–75, 138–139
independent variable, 53, 55, 137
information redundancy, 110–111
interaction effects, 196–200
internal experimentation team, 251–252
interrupted time series, 3
interviewing new hires, 241–244
Interviewing Users (Portigal), 34
Inverted Pyramid, 115

J

Jakob's Law, 104
Jarrett, Caroline, 34
JavaScript event tracking, 59
job interviews, 241–244
Johnson, Frank, 114–115
Johnson Box, 114–115

K

Kanban, in Conversion Design, 84–85
Kotter, John, 260–261
Krug, Steve, 105
Kübler-Ross, Elisabeth, 212
Kübler-Ross Change Curve, 212–213

L

lagging indicators, 63, 231
Landscape of User Research Methods, 24–25
language
　in company culture, 247
　compelling words, 127
　creating accessible plain language, 94–96
　localization, 96–98
　using data to craft content, 100–101
Laplace, Pierre-Simon, 168, 180
Laplacian curve, 180
Leading Change (Kotter), 260
leading indicators, 62–63, 137
learning budget, 248
Lentze, Jorden, 120–121
line charts, 194–195
linear thinking, 8, 10, 14–15
links and buttons, descriptive, 102
live chat limitations, 224
local maximum, 73–74
localization, 96–98, 229
logic sharing, in trust triangle, 245
looking and peeking, in analyze phase, 171–173

M

Manjunath, Shiva, 51–52
market foundations, as experiment ideas, 70
Martin, Dave, 247–249
matrix organized companies, 240
maturity spectrum. *See* Experimentation Cultural Maturity spectrum
maximum efficiency workflow, 78–79
maximums, local and global, 73–74
measuring success, 62–63
metric interactions, 196, 200
Metrics & Math, 133–135, 273
micro cultures, in company, 248
micro surveys, 30, 32–34
milestone celebrations, 246
Miller, Evan, 143
minimum detectable effect (MDE), 134, 138–139, 143–145
Mintzberg, Henry, 66
misdirection, 221
mission of company, 246
de Moivre, Abraham, 180
money, and test participants, 35
mores and norms, 236–237, 238, 247–248
motivating people in create phase, 114–128
　compelling words, 127
　encouragement, 128
　error messages and dead ends, 122–125
　feedback, 116–117
　framing techniques, 126
　good sequencing, 127–128
　Johnson Box and Inverted Pyramid, 114–115
　nudging prompt, 116
　page load speed, 118–121
motivation, as experiment ideas, 71
muscle memory, 213

N

navigation bars, 110–111
Nielsen, Jakob, 104
NN/g (Nielsen Norman Group), 105
noise in data, 229
non-inferiority bug fix, 169–171
non-inferiority tests, 163, 206
normal distribution, 180
norms and mores, 236–237, 238, 247–248
novelty effect, 212
null hypothesis, 47–49, 50
　testing, 45
Null Hypothesis Reminder, 132–134, 273
"number of nights," 106–107

O

objectivity techniques, 218–220
observational study, 4–5
Ohno, Taiichi, 84
O'Malley, Deborah, 111
On Death and Dying (Kübler-Ross), 212
"on" tracking, 59, 229–230
O'Neill, Sean, 214–217
optimal decisions, 63, 220
organizational psychology, and company culture, 260–261
outsourced experimentation, 250
overconfidence, 208

P

p-hacking, 207
p-values (probability values), 175, 186, 187–189, 190
page load speed, 118–121
peeking and looking, in analyze phase, 171–173
performance, as experiment ideas, 70–71
performance guardrail metrics, 120–121
personalities and role models in company, 237–238, 247
phone service mazes, 223, 224
photo galleries, and reviews, 108
plus-minus symbols, 195
pointillist paintings, 36, 40
Portigal, Steve, 34
positive and negative in error matrix
　detection of an effect, 151–158
　direction of an effect, 156–158
positively negative results, 227–228
post-task completion surveys, 33
power dynamics within company, 240–241
The Power of Minds at Work (Albrecht), 204
predicted value of an effect, 150–158
prediction section, 136, 270–271
prioritize phase, 67–89
　buckets of good experiment ideas, 69–71
　building the right thing: product-market fit, 68–69
　Impact Effort Matrix, 71–75
　Kanban tracking, 84–85
　running A/A tests, 76
　as step in Conversion Design, 18
　workflow design, 77–79
　workflow optimization, 87–88
　workflow problems: Four Bs, 79–83
probability, 177–178

Problem Statement Driven Hypothesis Blueprint, 51
procedural memory, 213
product descriptions, and review placement, 108
product development, digital, compared with human interaction, 22
product features list, with icons, 110
product foundations, as experiment ideas, 70
product images, 111–113
product-market fit, prioritizing to build the right thing, 68–69
product ratings, and sample size, 140
product use, in research, 25
Product Value Chains, 247–249
productivity, in workflow, 77
promoters, users as, 29
push-and-pull process, 84

Q

qualitative research, 23–24, 25, 26, 30
　in experience landscape, 38–40
　and optimal decisions, 220
quantitative research, 24, 25, 26, 30
　in experience landscape, 38–40
　and optimal decisions, 220
Quesenbery, Whitney, 71
questions
　answering along the way, 109–110
　for interviewing new hires, 241–244
　open and closed, in surveys, 30
　research, 23–24, 31
　trick, deceptive patterns, 222
quitting, strategically, 227

R

randomization
　defined, 54
　and sample ratio mismatches (SRMs), 57–61
randomized controlled experiment, 4–5, 6–8, 54–55, 165
range of results, 184
rating and review sites, 29–30
"read more" links, 102
real value of an effect (reality), 150–158
Reddit, 30
redundancy, 110–113
relationship-based research questions, 24
relative change calculations, 146–148
reporting analytics, 27

research environments, 26–27
research methods, 24–25
research questions, 23–24, 31. *See also* surveys
research strategy, 22–27
    environment, 26–27
    exploring data, 27–31
    exploring the real world, 31–41
    methods, 24–25
    questions, 23–24
return rate, as dependent variable, 52–55
review, rating, and reputation sites, 29–30
review boards, 219
reviews, 108, 116–117
risk mitigation, 225–226
Roach, William, 44
roach motel, 221
Rohn, Jim, 202
Rohrer, Christian, 24–25
role models in company, 237–238, 247
rule of thumb, 159
run time, in testing, calculating, 135, 140–149, 162

S

Saffer II, HM, 36
sales teams, and research, 35
sample diversity, 140
Sample Ratio Mismatch Checker, 58–59, 60
sample ratio mismatches (SRMs), 57–61, 172
sample size, in testing, calculating, 135, 140–149
sampling distribution, 174–175, 176–179, 182–183
scarcity, and decision making, 210–211
science, as discipline of evidence, 3–4
scientific method, 45–46
screen readers, 102
search engine optimization (SEO), 102, 113
seasonality, 162
secondary needs, 83, 88
selection bias, 213
sequencing, for managing tasks, 127–128
service-level agreements (SLAs), 121
shadowing, 21–22, 34
significance level, 135, 159–160, 164–165, 187
significant results, 188, 194–195
Silenski-Cahill, Lucas, 48
sizing information, as independent variable, 50, 53–55
Sketch, 75
sneaky add-ons, 222
social norms and mores, 236–237, 238, 247–248

software updates, 230
spaghetti testing, 207
speed vs. velocity, 77
Spool, Jared, 11
Statistical Methods for Research Workers (Fisher), 45
statistical power, 135, 159–161, 164–165
statistical significance, 160, 187, 188–189
statistics, birth of modern, with tea test, 43–45
Stephens, Jonathan, 106–107
Stone, Martin, 86
strategic quitting, 227
Substack, 30
"A Sunset Path" (Saffer), 36
superiority tests, 162–163, 206
surveys, 30, 32–34
Surveys That Work (Jarrett), 34
swipeable carousels, 230
systematic reviews, 4–5, 219–220, 232
systems thinking, 8–10

T

task efficiency, 78–79
tea test, birth of modern statistics, 43–45, 47
technical stability dial-up, 214–217
technical stuff, causing good ideas to fail, 228
Tesco, 214
test phase, 131–167
    confusion matrix (error matrix), 150–158
    goal selection, 134, 135–137
    Metrics & Math, 133–135
    minimum detectable effect (MDE), 134, 138–139, 143–145
    Null Hypothesis Reminder, 132–134
    run time calculations, 135, 140–149, 162
    sample size calculations, 135, 140–149
    statistical power and significance level, 135, 159–161, 164–165
    as step in Conversion Design, 18
    test type, 162–163
    validate template: test like you're wrong, 272–273
test types (superiority and non-inferiority), 162–163, 206
Texas Sharpshooter fallacy, 205–206
text highlight colors, 103–105
three-bin stocking process, 84
time, between tech evolution and user behavior, 230
tipping, 21
tooltips, 109–110

toothbrush experiment, 48–49
tracking, 59, 229–230
trade-offs, weighing, 220
traditions of company, 237, 238, 246–247
traffic interactions, 198–200
transcreation, of language, 98
translation, of language, 96–97
transliteration, of language, 96–97
treatment, in testing, 6–7, 132
trick questions, 222
true and false in error matrix, 150–158
true negative, 153, 158
true positive, 153, 156–157
trust triangle, 244–246
Twyman, Tony, 218
Twyman's Law, 218
Type I error. *See* false positive
Type II error. *See* false negative

U

underpowered results, 191
understand phase, 21–41
   curiosity as research, 22–23
   exploring data, 27–31
   exploring the real world, 31–41
   research environments, 26–27
   research methods, 24–26
   research questions, 23–24
   as step in Conversion Design, 18
undesired defaults, 222
usability, 71, 92–96
usability bugs, 223
usability tests, 34
user behavior, in Product Value Chain, 257–259
user journey maps, 36
user skew, 217
users, promoters and detractors, 30

V

validate template, 132, 272–273
value creation, in Conversion Design, 10–13
value cycle, 12–13, 266
value of an effect, real and predicted, 150–158
values of company, 246
vanity metric, 62
variables, in experiments, 52–55
variance, of data, 173–174
variant, in testing, 6–7, 132
Vermeer, Lukas, 58–59, 164–165, 196–200
Vickers, Andrew, 180
vision of company, 246
visitors, 189
visual signals/cards (Kanban), 84
visualizations of results, 189–195
   box plots, 194
   checking results, 191–192
   line charts, 194–195
   plus-minus symbols, 195

W

Web Content Accessibility Guidelines (WCAG), 71
A Web for Everyone (Horton and Quesenbery), 71
weighing trade-offs, 220
Wesseling, Ton, 249
What Is a P-Value Anyway? (Vickers), 180
win-win decisions, 220
worker efficiency, 78–79
workflow
   design, 77–79
   optimization, 87–88
   problems: Four Bs, 79–83

Y

YouTube, 30

Acknowledgments

Although my name is on the cover of this book, its creation was *truly* a team effort. From the moment I started writing, I called it "ours." And so many of you have stepped up to contribute your strengths and support to this final product. This is a non-exhaustive list of "thank yous" to people who've really made an impact on the quality of this book:

My dad, George Weigel, with whom I've had countless hours' worth of video calls. Thank you for helping me get the content structure and flow right, as well as generally being my writing coach through the years.

My mama, Susan Weigel, who loved me most by letting me go (to a foreign country by myself when I was just a teenager). I've chased my dreams because you gave me the confidence to step out into the world and discover new things.

Lucas Bernardi for running your "Statistics Panic Attack Hotline," and for sitting in the library and on video calls for hours on end with me. You ensured the math and experimental design content was accurate by graciously pointing out all the stupid mistakes I made and then patiently explaining why I was wrong. This book would have been a total train wreck without you.

Francesca Catanuso—the best wife I've never had. Not only are you super smart and an experienced professional, but you're also an *exceptionally* caring friend. Thank you for all the content consults, encouragement hugs, and home-cooked meals you made for my family as I worked toward my deadlines. You're a stellar human, and I'm proud to call you my friend.

Shiva Manjunath for making me a "Certified Homie." This honor came with the incredible perk of text message access to get feedback from you at literally all hours of the day and night. Your agency-side perspective was invaluable to making this book relevant to a broader audience than just folks who work in big tech with hundreds of millions of users. I appreciate you and all of your memes.

Britta Schumann for helping me get off the struggle bus that was Chapter 5. Your perspective and thoughtful feedback was imperative to help get me over that hurdle. Now that I'm done with this, let's go fix the roof and plant some cucumbers.

Dewi Williams and Alexis Oh for not only reviewing my book and giving feedback, but also being my designer sparring partners and helping me with the illustrations. That brings me to **Serbastian Tan and Martin Pavlovic**, too. The willingness all of you showed to hop in my hot-mess of a Figma file and draw graphics made that part of the process a lot more fun and manageable. I'm so grateful for your skills and your generosity.

Stuart Frisby and Lukas Vermeer for always entertaining my antics. I've learned so much from both of you—but don't let that go to your heads. Thanks for writing the foreword and lending me some of your street cred. Dinner's on me.

Els Aerts for cheering me on and sharing your research expertise. I'm so lucky to have you in my court. Els basically wrote Chapter 2 and gave me feedback on my conference talks over the years. She's been a role model as a top-notch conference speaker, and she *always* makes space for me and other women to come take the stage.

Lou Rosenfeld for taking a chance on me. It has been a dream come true to orbit in your circle of authors, many of whom I've looked up to when I was a baby designer. Your books and dedication to progressing our industry has enriched my career and life. You are a shining example of humanity at its finest.

Marta Justak for being my coach and my rock throughout this entire process. You are a true leader and the secret sauce behind the Rosenfeld brand. Your encouragement kept me going and your advice kept me sane. I'm grateful I had the opportunity to work with and grow from your mentorship.

To the rest of you in no specific order:

Anna Popova, Martin Stone, Dave Martin, Sean O'Neill, Nekeia Boone, Jonathan Stephens, Ben Labay, Peep Laja, André Morys, Kevin Anderson, Ton Wesseling, Dr. Danielle Guzman-Orth, Cambria Davies, Alex Birkett, Positive John, Jonas Alves, Ruben de Boer, Ron Kohavi, Michael Aagaard, Deborah O'Malley, Christian Rohrer, Dr. BJ Fogg, Rebekah Hickey, Steffan Williams, Tatsiana Khiliutka, Alexander Richter, Ed Roberts, Matt LeMay, Jorden Lentze, John Melia, Preston Daniel, Pieter Boonstra, Lisa Carter, Beth Archibald Martin, Dr. Judith Yaaqoubi, Jason Clauss, Adam Nassr, Eric Reichbaum, and many, many more.

I know it's highly likely (my statistics knowledge in action!) that I'm forgetting people. So, if that's you, please forgive me.

And to *you*, dear reader. Thank you for trusting me with your time and attention. I hope you found some of what I had to share valuable. Now, go out into the world, test stuff, and *make things better!*

Rosenfeld

Dear Reader,

Thanks very much for purchasing this book. There's a story behind it and every product we create at Rosenfeld Media.

Since the early 1990s, I've been a User Experience consultant, conference presenter, workshop instructor, and author. (I'm probably best-known for having cowritten *Information Architecture for the Web and Beyond*.) In each of these roles, I've been frustrated by the missed opportunities to apply UX principles and practices.

I started Rosenfeld Media in 2005 with the goal of publishing books whose design and development showed that a publisher could practice what it preached. Since then, we've expanded into producing industry-leading conferences and workshops. In all cases, UX has helped us create better, more successful products—just as you would expect. From employing user research to drive the design of our books and conference programs, to working closely with our conference speakers on their talks, to caring deeply about customer service, we practice what we preach every day.

Please visit **rosenfeldmedia.com** to learn more about our **conferences**, **workshops**, **free communities**, and **other great resources** that we've made for you. And send your ideas, suggestions, and concerns my way: louis@rosenfeldmedia.com

I'd love to hear from you, and I hope you enjoy the book!

Lou Rosenfeld,
Publisher

RECENT TITLES FROM ROSENFELD MEDIA

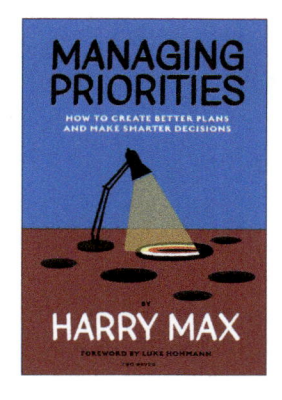

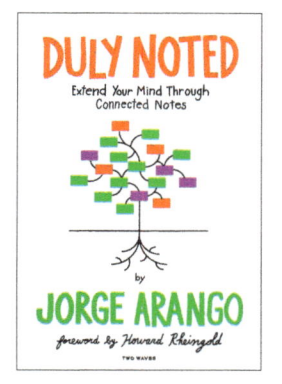

Get a great discount on a Rosenfeld Media book:
visit **rfld.me/deal** to learn more.

SELECTED TITLES FROM ROSENFELD MEDIA

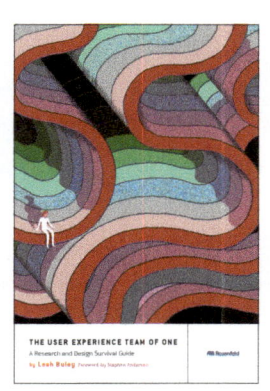

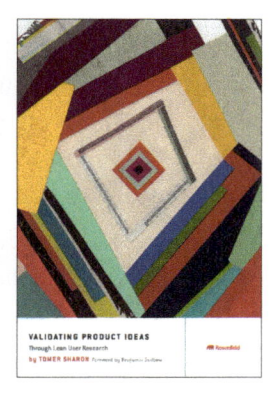

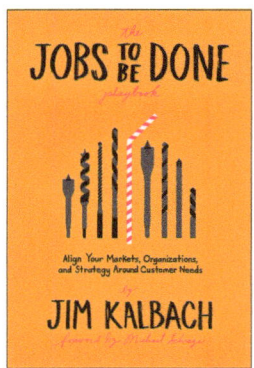

View our full catalog at **rosenfeldmedia.com/books**

About the Author

ERIN WEIGEL is a principal designer and senior design manager. She has a Bachelor's degree in fine art painting from The Maryland Institute College of Art (MICA), and she has studied at Konstfack University College of Art & Design in Stockholm, Sweden.

Her career began with waiting tables and working retail. She eventually worked her way up to Principal Designer at Booking.com (BKNG), the world's largest online travel company. She's worked in *many* different roles within product teams. She's spent time as a product manager of data science, a developer building websites and coding up emails, and she is currently a Senior Design Manager at Deliveroo.

While at Booking.com, Erin designed and coded more than 1,200 A/B tests. She analyzed and learned from thousands more. Her unique approach to product design and development combines a service mindset, customer-centric thinking, and A/B testing to deliver highly converting products that customers love.

Nowadays, she shares what she learns about design and experimentation at conferences and workshops around the world. You can follow her on LinkedIn at **www.LinkedIn.com/in/ErinDoesThings** or visit her personal website at **www.ErinDoesThings.com** for more information.